Wakefield Press

ELEPHANTS AND EGOTISTS

Patricia Sumerling is co-author of the landmark publication *Heritage of the City of Adelaide: An Illustrated Guide,* 1990, and author of *The Adelaide Park Lands: A social history,* 2011. She debuted as a novelist in 2010 with a historical crime mystery *The Noon Lady of Towitta.*

By the same author

Heritage of the City of Adelaide: An Illustrated Guide
The Adelaide Park Lands: A social history
The Noon Lady of Towitta

Elephants and Egotists

In Search of Samorn of the Adelaide Zoo

PATRICIA SUMERLING

Wakefield Press

Wakefield Press
16 Rose Street
Mile End
South Australia 5031
www.wakefieldpress.com.au

First published 2016

Cover designed by Liz Nicholson, designBITE
Text designed and typeset by Wakefield Press
Printed in Australia by Griffin Digital, Adelaide

National Library of Australia Cataloguing-in-Publication entry

Creator:	Sumerling, Patricia, 1945– , author.
Title:	Elephants and egotists: in search of Samorn of the Adelaide Zoo / Patricia Sumerling.
ISBN:	978 1 74305 399 7 (paperback).
Notes:	Includes bibliographical references and index.
Subjects:	Samorn (Elephant).
	Elephants – South Australia – History.
	Elephants – Southeast Asia.
	Endangered species.
	Zoos – South Australia – Adelaide – History.
Dewey Number:	599.61

CORIOLE
McLAREN VALE

'You know … they say an elephant never forgets. What they don't tell you is, you never forget an elephant.'

Actor Bill Murray from the film *Larger than Life*

CONTENTS

FOREWORD

Leo Oosterweghel

I knew Samorn during the late 1970s when I was a bird keeper at Adelaide Zoo. She was a feisty girl who was very particular about whom she liked, and very clear about those she had no time for. Fortunately for me, we were friends, and I would greet her every day on my way to work.

Over the years much has changed about the way zoos keep elephants. In the past there was very little understanding about what a zoo elephant needed in order to be content. They were often kept like domestic livestock, housed in stalls overnight and without the social contact we now know is vital to their welfare. Many were kept alone, or teamed up with elephants they didn't know. They became frustrated and developed unpredictable behaviours, and were then labelled 'difficult' or 'aggressive'.

Now, thankfully, we are a little wiser. We know that the inspiration for how to keep elephants in zoos must come from the wild, that elephants are highly intelligent and social animals, and that being part of the herd is the key to their wellbeing.

At Dublin Zoo we have a small herd of Asian elephants, all related. They live very natural lives, eating, sleeping and exploring with minimal human interference. In 2014 three calves were born into the family, and we now have eight hungry mouths to feed.

Having read Patricia's book, which I found well researched and fascinating, I would recommend it to anyone with an interest in history and in the way we used to look after animals. I know that Samorn would give her stamp of approval!

Leo Oosterweghel
Director Dublin Zoo
Ireland

1

Death of an elephant

ON 10 OCTOBER 1994 the zoo's silver bus, filled with visitors, pulled up at the boundary of the animal enclosure at Monarto Zoo where Samorn the Thai elephant (known as 'Sam' or 'Sammy' by those who worked with her) was always waiting to greet her visitors. She was supposed to be there as usual, but instead of that friendly craggy head and the trunk writhing to show her pleasure at seeing them, the visitors were met by a most peculiar sight. All they could see were four large dark grey legs sticking straight up from the dry moat. This was the ha-ha boundary designed and constructed to keep Samorn and other large animals secure from wandering off.

The passengers, when they realised what they were looking at, cried the obvious. 'Look there's an elephant upside down!' 'Why is she upside down, Mummy?' 'Why doesn't she roll over and get up?' 'Is she all right?' Or words to that effect.

The panic-stricken driver, not believing what he thought he'd seen, scrambled from the bus and rushed to look into the moat where Samorn was trapped. He could see that she had fallen or slipped into the moat, rolled upside down and lodged herself so securely that escape was impossible. Worse, he could not tell if she was alive or dead, for there was no movement or flicker of life when he called 'Sam'. And he couldn't get to touch her because, of course, the boundary ha-ha was designed so no one from the visitor side of the moat could climb into the enclosure.

He clambered back on the bus and spoke into his radio telephone to ask the business centre for immediate assistance. And then he turned to his bus full of distressed visitors and said something like, 'It's going to be all right, kids. And, ladies and gentlemen, please don't be alarmed. I've rung for help and we'll be able to get her out in no time as the zoo has wonderful equipment for matters such as this. She's just slipped into the moat in such a way that she can't move without assistance. I'll keep you informed as I hear how the rescue goes. I'm sure there's no need for alarm. She's a tough old girl. I don't want you to be more upset than you are so we'll go and see the other animals and come back when Samorn's on her feet.'

All the children, greatly relieved, cheered and clapped the driver as he started the bus engine.

And so that morning Samorn's keeper, Paul O'Donoghue, was in the moat with her for a couple of hours, talking to her and pacifying her in her extraordinary situation while a crane was preparing to rescue

Samorn in the late 1970s. (Photographer Chris Thompson)

her. Both zoo vets were called and arrived as soon as possible. Sadly, minutes after the crane lifted her out of the moat and placed her gently down with her legs tucked beneath her, her life quietly ebbed away.

Samorn was Adelaide's fourth zoo elephant. For the time being there would be no more elephants at the Adelaide Zoo or at its Monarto Zoological Park, even though there had been plans for several more to join Samorn.

Unfortunately, ever since zoos have done away with bars and fences and dug moats instead, elephants dying in them is not an uncommon occurrence. The fact that the bottom of the moat at Monarto had not been rounded off so that a trapped animal could easily free itself, did not help Samorn's plight.

While quite a few elephants have slipped into moats accidentally, at Sydney's Taronga Park in May 1949, a small girl in tears told attendants that she had watched one elephant bump another elephant named Nellie, a 26-year-old several-tonner, into a deep moat. Like Samorn she was found on her back wedged with her feet in the air. Attendants with blocks and tackle battled for three hours to try and hoist her into a sitting position, but she died before they could rescue her.

Several years before in the Melbourne Zoo, Peggy the elephant playfully pushed Betty her companion into the moat, again creating the problem of how to get her out. In her case, it was a success story, for bales of tightly packed straw were arranged to create a slope for her to walk up. Furthermore, both elephants learned from this incident for they were regularly seen afterwards walking down the sloping side of the moat and then climbing out again.

Of course, when the news was spread that Samorn had died at Monarto there was an outpouring of grief from many of the grown-up children who'd had their own special time with her during more than three decades. But Samorn's death coincided with the many changing attitudes of the time about keeping wild animals, particularly elephants, in zoos for the express purpose of display and the amusement of fee-paying customers. Even breeding programs within zoos for some animals were being questioned.

Zoos all around the world were amending their charters as animal conservation and the harsh reality of endangered species challenged their relevance. There is no world human population peak in sight, and seemingly few serious initiatives to bring about planned reduction (though one mustn't overlook the draconian measures of China's one-child policy, and the worldwide family planning foundation set up by Bill and Melissa Gates in July 2012). As the world human population soars, the demand for virgin land grows unabated, destroying natural habitats of such endangered species as orang-utans, pandas, tigers, elephants, Syrian bears, Madagascar aye-ayes or, in Australia, Carnaby's black cockatoo from the southern parts of Western Australia. Even more elusive is the legendary termite-mound-nesting paradise parrot of southern Queensland. Its supposed extinction in the 1920s is yet to be proven for it is still listed in bird books as the rarest bird in Australia. As ornithologist Penny Olsen states, 'After all how can extinction be proven?'

As numbers of endangered species climb, and zoos around the world collaborate for better solutions by keeping details of the gene pools of their own captured or zoo-raised animals, the importance of safeguarding natural habitats and provision of corridors linking areas is pursued in earnest. In C.F.H. Jenkins' book of 1977, *Noah's Ark Syndrome*, he states, 'International and interstate liaisons have resulted in stud books for many rare animals and inter-zoo loans and exchanges are now common in a bid to promote breeding and thereby to preserve endangered species'.

The glorious and quirky Elephants on Parade sculpture show of over 250 baby-sized painted and adorned elephants was on public display in London for over two months in 2010. It was then auctioned and raised over £4 million to buy tracts of land in Assam, northern India, for use as corridors from one area to another for its local elephants, thus demonstrating what is possible. Acquiring rainforests in Sarawak to guarantee homes for orang-utans continues against impossible odds, while reserves created for the critically endangered rhinoceros in Java had less than 40 rhinos in 2012 because of poachers. Perhaps we should be reminded of Penny Olsen's quote about extinct creatures.

When the tiny brown and orange blotched grassland copper butterfly (*Lucia limbaria*) was spotted by Janet Subiago in the middle of Victoria Park in Adelaide in 2011, she and a dedicated group of conservationists successfully fought government authorities for its protection. For years thought to be extinct, its survival is due to a small acreage of remnant grasses not tampered with for more than 165 years – even while the vicinity was used for horse-racing. It is one small triumph amid an endless list of conservation battles fought but not always won.

2

Early South Australian elephants, the Adelaide Zoo and menageries

THE SMALLEST CITY ZOO in Australia, the Adelaide Zoological Gardens of the Royal Zoological Society of South Australia, is an eight-hectare (20-acre) wooded and shady area carved out of a small corner of the Botanic Park, which is managed by the Botanic Gardens. They in turn are part of the city's encircling Park Lands, managed by the Adelaide City Council and the newly formed Park Lands Authority. This green verge of parklands swaddles the small city, of only about 421.5 hectares (1042 acres), like a huge figure of eight around its two separate parts that straddle a creek, grandiosely called a river (the River Torrens). Indeed, a boutique zoo for a boutique city.

The enviable Mediterranean climate of Adelaide that is much like Morocco and Tunisia ensured that animal exhibits on show from the mid-nineteenth century generally thrived and sometimes multiplied. However

1898 plan of Adelaide Zoological Gardens.

there was great ignorance about and neglect in caring properly for captured wild animals in times past, no matter how mild the climate.

Throughout the nineteenth century there was a healthy tradition in Australia for displays of exotic animals. As seamen passed through South Australia or jumped ship to escape poor shipboard conditions, they often sold or left their pets in the pubs in exchange for food, board or other pressing debts. While birds, snakes, other reptiles, and monkeys were the most popular species, unusually at the Union Hotel in Waymouth Street in May 1849, a 'black tiger' (most likely a jaguar or leopard) was exhibited for a short period and caused much excitement for its size and 'wild' behaviour. It had arrived by the ship *Cacique* from Malacca.

The *Adelaide Times* for 21 May 1849 declared that the 'Black Tiger is now exhibited by Mr Creech at the Union Inn'. As you would expect of an animal in a thoroughly wild state when approached, it roared, spat and showed many 'curlings of the tail in true jungle style'. It measured from six to seven feet from mouth to tip of tail and was noted for its beautiful sleek fur. Because it was so rare to see such an animal in South Australia, it was expected that crowds would flock to see it for themselves.

The newspaper speculated that Mr Creech would probably find the animal an expensive lodger. As it turned out, feeding the 'tiger' was rather taxing and it caused plenty of trouble for its owner who ended up in court accused of stealing his neighbour's ducks to feed it. When the case was thrown out through lack of reliable evidence, Mr Creech announced that he was shipping his pet to London. In fact it ended up at the Ship Inn in Hobart by November of that year.

Before this, in 1838, the second governor of South Australia, George Gawler, kept a menagerie of animals for his children at Government House. Gawler's daughter Julia wrote in her diary: 'We are fencing an aviary. I have a cockatoo, owl, & laughing Jackass given to me. Our farm increases every week. 20 goslings, turkey, a hen, 4 cows, 5 calves, a young kangaroo and a Newfoundland dog called Rose.'

In July 1840 Gawler's son John wrote to his Aunt Jane: 'We have quite a farmyard now, 5 pigs with young ones … We have plenty of poultry, 2 hens sitting, we sometimes find 20 eggs a day. We also have 4 Spanish geese … and five English … 2 ducks, a drake, 1 muscovy duck, 7 cows, 6 calves and 2 emus.' He described the emus as very tame, although the young one occasionally wandered into town, 'where it is chased from one end of the street to the other by packs of dogs, but it always returns safe home …' The emu's antics occasionally reached the local newspapers.

Within a short time following the Botanic Gardens' public opening in October 1857, a growing number of birds and animals were put on display in a small menagerie 'in a variety of wired, fenced and barred enclosures as established by George Francis'. When an American bear was donated a nearby notice informed the public that:

> Mr. Bruin, the bear, who lately came here as an assisted immigrant, being delighted with the climate … intends to remain here for some time – at least till the squatting question is decided … He is now having a house built for him … He has no objection to visitors poking fun at him, but has the strongest objection to the poking of sticks and parasols.

After Bruin was put on display in 1864 the *South Australian Register* for 19 September announced that as many as 2,720 persons visited the gardens one Sunday afternoon to see him.

The 19th century was a time of minds inquiring of all things scientific as explorers and travellers returned from their grand tours or explorations with exotic flora and fauna that encouraged the establishment of gardens and zoos in major progressive towns of the Western world. Cashing in on the curiosity to see exotic animals, pleasure gardens were established internationally that sometimes included small menageries. Sadly, through lack of funds, but mostly through ignorance and neglect about keeping exotic exhibits alive, such places survived but briefly. When animals were not rescued by wealthy benefactors they were usually destroyed rather then letting them starve to death through lack of funds to feed them.

It was during this mid-Victorian era of curiosity and 60 years after European settlement of Sydney in New South Wales, that two baby Indian elephants were put aboard the ship *Royal Saxon* in Calcutta bound for the Australian colonies in 1851.

These two little elephants were the very first to arrive in Australia. They were just 20 months old when they headed for Hobart where one, called Rajah, was off-loaded. The other one, bound for Sydney, was going into the care of the publicans William Beaumont and James Waller who owned the Joseph Banks Hotel where they had established their Botany Bay Pleasure Gardens in about 1849. The Sydney gardens were copied from the famous London Cremorne Gardens on the banks of the Thames at Chelsea.

The elephant called Rajah that landed at Hobart spent some weeks there before he was shipped to Melbourne where he lived for several months earning his keep by being put on display to the public. Then in July 1852 Rajah, now called Jumbo, was placed aboard the ship *Louisa* bound for Port Adelaide. When he arrived as

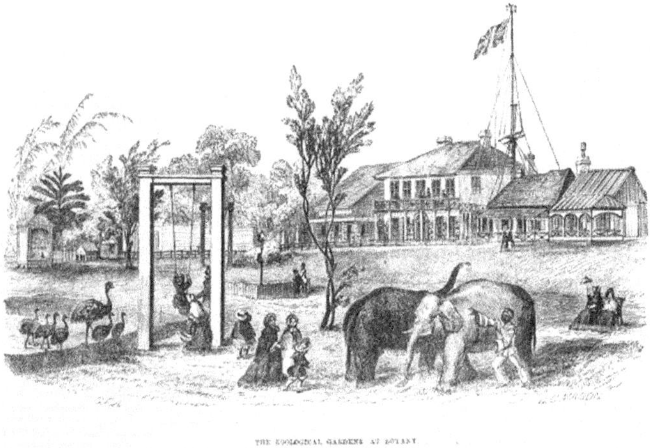

Two elephants at the Botany Bay Pleasure Gardens, Sydney.
(Illustrated Sydney News, *30 June 1855)*

South Australia's first elephant, he was walked the 13 kilometres (eight miles) to Adelaide and put on display for an entrance fee at the Edinburgh Castle Hotel in Currie Street (which is now the oldest trading hotel in South Australia).

Like the other little elephant destined for the Sydney pleasure gardens, the Adelaide-bound one then lived at a 'Cremorne pleasure gardens', this one managed by Thomas Bentley, and which included a menagerie established alongside his pub, the Cremorne Hotel on Unley Road. After Bentley obtained a licence for his hotel in 1852, Jumbo was the Cremorne Gardens' star attraction, with his exploits sometimes recorded by the newspapers.

But the menagerie, technically South Australia's first zoo open to the public, unexpectedly ceased operations

Rajah lived here from 1852 to 1855. Frearson's Weekly, *22 November 1879. (State Library of South Australia)*

nearly three years later. Mr Bentley, who had figured in a case under the *Masters & Servants Act* in the Supreme Court for failing to pay wages to the Indian mahout hired to look after Jumbo, went broke. After he declared his insolvency in court in July 1855, his predicament was formally recorded in the local government gazette.

In the fiasco, the pub and the entire menagerie of animals, including Jumbo, were put up for auction. In a rash moment, Jumbo was bid for by John Smith of the Smithfield Hotel who used him for ploughing tasks. As Smith soon found out, two bullocks could actually work faster than Jumbo and several months later he was sold to Charles Matthews of the Gepps Cross Hotel, who re-named him Tommy. Matthews owned Tommy until the creature's early death after catching a chill in 1858. He was only eight years old. His owners weren't informed about keeping elephants; although they claimed to have cared for his well-being, it could be argued that his death may well have been avoided had there been more knowledge about the proper care of elephants.

With them needing at least 100 kilograms of food a day, depending on size, age and type of elephant, endless amounts of bread, cakes and beer from the pub's bar could hardly be called the right tucker for a growing elephant like Tommy. In six years in Australia, this young elephant had figured in three court cases, taken part in foot races and parades, been put on public display in three colonies for a fee, had three names, and travelled on three ships. and even had a walk-on part on 19 April 1855 at Adelaide's oldest theatre, then called the Royal Victoria Theatre, located in Gilles Arcade, off Currie Street.

It was more than 20 years after Tommy's death before another elephant resided in Adelaide. This came about after the South Australian Acclimatization Society was formed in 1878 after several abortive attempts to form a zoo from the 1860s onwards. However, in time for the zoo's public opening in April 1883, the society was registered with the name of the South Australian Zoological and Acclimatization Society. And through the generosity of the wealthy entrepreneur and board member, Sir Thomas Elder, it received in that year the gift of a Siamese elephant, appropriately called Miss Siam. She became the zoo's first animal to have a name, recalling London Zoo's Jumbo who was reputedly the first named zoo animal in the world.

Naturally, because of Miss Siam's awesome size and her approachability as viewed through the eyes of a small child, she became the zoo's favourite beast and the colony's first 'public pet'. In slightly more than 100 years, several generations of children have grown up with a series of just five elephants to love throughout their childhood, so that in their later lives they could whimsically recall the day they had their own special interaction with the zoo elephant, be it Miss Siam, Mary Anne, Lillian, Samorn or Tara.

This book is the story of one of these elephants, Samorn, the Adelaide Zoo's fourth, who as the 'state's pet', delighted generations of children and their parents between 1956 and her sad death in October 1994. Although Samorn lived with an array of animals, zoos are as much about the people associated with them as the animals themselves. Such people include elephant keepers, zoo presidents and directors, animal traders, big game hunters, wildlife smugglers, and the throngs of happy children and grown-up children who flocked to see Samorn. Then there are the other elephants that came before her.

There are now generations of zoo visitors with their own special tale about Samorn lingering in their fond memories of childhood, the oldest of whom are now in their 70s.

3

The elephants of Adelaide Zoo

Miss Siam

The first elephant at the Adelaide Zoo, Miss Siam, was a gift from Sir Thomas Elder who had organised for her to be shipped from the King of Siam (now known as Thailand) in November 1883. She was brought to South Australia by the Adelaide Steamship Company at no expense to the zoo, and the South Australian Company generously gave free use of wharf and crane at Port Adelaide when she arrived. Miss Siam had travelled to Australia on the steamer *South Australian* with an elephant destined for the Melbourne Zoo. Mr Meagher, an elephant keeper from the Melbourne Zoo, cared for both animals on the voyage. When Miss Siam arrived at Port Adelaide the large crate in which she had travelled was placed onto a trolley and carted to the Adelaide Zoo with her. After she had settled in, 'very docile' Miss Siam was not at all upset by the huge attention paid to her by multitudes of visitors, including those just wanting a closer examination of her and her extraordinary trunk.

Director Richard Ernest Minchin met King Chulalongkorn of Siam in 1886.

The arrival of a first elephant to the zoo as the *attractio maxima* quickly boosted the number of visitors. They were inquisitive to see what the largest land animal in the world was really like in the flesh, even though she was not yet half grown. When she grew bigger she was put to work with a howdah made to fit her back. She carried 3150 children in her first six months, which at three pence a ride raised a little under £40.

She was in the company of 48 species of mammals, 66 species of birds and seven species of reptiles. As the most popular denizen of the Adelaide Zoo, when Miss Siam outgrew her wooden cage a new ornamental elephant house more befitting her status (with 1900 etched into the stucco) was built for her. Constructed of concrete it was designed like an oriental pavilion crowned by a dome that reached 36 feet above the ground, and decorated by a smaller cupola at each corner of its four sides. Of a generous size to accommodate one elephant, its architecture symbolised the

Miss Siam at the Elephant House, circa 1901.
(Royal Zoological Society of South Australia Inc.)

great British Empire that was then at war with the Boers in South Africa. Zoo buildings such as these were very much a statement about the power of empires and their colonies that collected exotic species from the far-flung corners of the world.

With Miss Siam being the first elephant at the zoo, it is unlikely that any of the zookeepers had experience looking after such creatures. So it is of no surprise to find that within Miss Siam's first year, one of them advertised through the *Advertiser* for 7 November 1884: 'Situation wanted by late Elephant-keeper, any capacity, thorough accountant, ride or drive, handy man; good references.' One can only wonder why he was looking for a new job. Perhaps he had found working with the elephant too stressful.

Miss Siam, who was at the zoo for 20 years, lived in the grand elephant house for less than five. With her typical voracious appetite, there were serious concerns for her rapidly deteriorating health when she went off her food for about a fortnight. Then on the day of her death on 7 May 1904, in the words of the director, the *Advertiser* quoted 'she lay down like a cow and went off in five minutes'. An autopsy of the 'children's friend' revealed an inflammation and suppuration of the liver caused by an obstruction of the bile ducts – in other words, it was an enormous gall stone. Weighing 21½ pounds it was larger than a man's head or the size of a football, and it was preserved.

The Chronicle, *14 May 1904.*
(State Library of South Australia)

Miss Siam on display at the South Australian Museum, circa 1905. (State Library of South Australia B 72463/137)

Miss Siam was not completely lost to the sorrowful juvenile population when she died. She was retained by the city's museum on North Terrace so that the skill of the taxidermist could make her look as much like the breathing original as possible. For over 110 years now, she has looked out of a glass display cabinet, with very poorly manicured toes for all to see. At the time of Miss Siam's death it was noted by the media: 'Wasn't it a coincidence that within the last six months, both the zoological gardens in Melbourne and Perth had also lost their only elephants?' The Melbourne elephant, it was claimed, was Miss Siam's sister who had travelled to Australia with her in 1883.

Because Miss Siam had been the number-one attraction, and one that could earn her entire keep, the zoo wanted to replace her as soon as possible, but a new elephant at that time would have cost at least £500 including the freight and other charges. When the second director, Alfred Corker Minchin, informed the media that it was a sum the zoo couldn't afford without help from the public, within a week he had collected £40 from generous donors. A collection box was then placed near the zoo gates and a free admission day was expected to raise more donations. Within three weeks anonymous donors along with others calling themselves 'Well wisher' or 'A child's friend' boosted the amount to £100.

When the zoo had raised about £200 it announced that another elephant had been procured from Ceylon (now called Sri Lanka) by John Hagenbeck who was on that island buying a number of elephants to be trained for exhibition purposes. (John was the brother of the famous Hamburg Tierpark Zoo owner, Carl Hagenbeck, who was credited with designing the first naturalistic zoos, replacing bars and fences with moats.) The Adelaide Zoo wanted to obtain an elephant of 'even temper and kindly disposition, who would not only have a handsome appearance, but would be susceptible of education in the performance of the duties required at the Gardens'.

Mary Ann

Mary Ann, the second Adelaide Zoo elephant, arrived as an eight-year-old from Ceylon in November 1904. She earned about £3 per day giving rides at three pence a time, earning more in a day than her keeper in a week. And to boot, her work was for just two-and-a-half hours every afternoon between 2 and 4.30 pm. By 1932 large numbers of visitors meant much needed revenue, for it was at the depth of the Great Depression. Money was tight yet she was estimated to have carried more than 6000 children per year in the howdah at three pence per ride, which was about the cost of her keep for the year.

Mary Ann and children from Estcourt House in 1923.
(State Library of South Australia PRG 280/1/40/210)

Although there were a couple of serious attacks by animals at the Adelaide Zoo (one which led to the death of Keeper Samuel May after his arm was torn off by a polar bear in February 1920) there have been surprisingly few documented. Unquestionably, there are many unreported escape stories that circulated among zookeepers. One escape that reached the media in January 1932 concerned 36-year-old Mary Ann after she attacked her keeper, Mr Bradshaw. Without warning she grabbed Bradshaw when being fed biscuits by a visitor. Winding her trunk about his neck was no playful gesture. He remarked that had he not realised in a split second what was about to happen and walloped her trunk to escape from its coils, nothing could have saved him from being lifted inside the caged enclosure and trampled to death. Chillingly, he recalled the feel of the coarse hairs on her trunk scraping his face.

Mary Ann had raised scares previously for she had also tried to clutch the ice-man in her trunk (he provided some animals with blocks of ice in the summertime), but he too managed to struggle free. The opinion was that it seemed out of character for Mary Ann who was usually very docile, and allowed the 'intimacy' of her nails to be manicured or her hide to be swabbed with oil.

Mary Ann died from peritonitis and enteritis, caused by a twisting of the bowel, on 19 January 1934 when she was about 38 years old. But because she had been 'the goose to lay the golden eggs', paying entirely for her keep through the rides she gave, no time was lost in finding a replacement. A cablegram was sent to the Mayfield Kennels in Singapore to ascertain whether there was a female Asiatic elephant for sale. Her cost was to be partly paid for by exchange of Australian animals.

Lillian

Lillian was the much-loved third elephant to reign at Adelaide Zoo. She was shipped to Adelaide as a four-year-old from Ceylon via the Singapore Zoological Gardens (aka the Mayfield Kennels) in May 1934. At 4 feet 9 inches (1.44 metres), she cost the zoo £165.

One year after her arrival the zoo decided that to determine her rate of growth she should be weighed annually. It became a minor city event for several years as she was walked through the city streets to the East End Market to be placed on the licensed weighbridge used for weighing fully laden lorries of fruit and vegetables.

In a country where blokes would make a bet on two flies crawling up a wall, guessing the weight of a several ton animal to the nearest pound was almost as ludicrous. So it was, in 1937, Lillian's third year at the zoo, that guessing the elephant's weight became a fund-raising event with cash prizes awarded.

Adelaide has had many a quirky club undertaking weird and curious activities, some for the purpose of raising funds for good causes while having fun in the process. One club active in the 1930s was the

Lillian's arrival in 1934.
(Royal Zoological Society of South Australia Inc.)

Taking Lillian to the East End Market for the annual weighing, circa 1937. (Royal Zoological Society of South Australia Inc.)

Liars' Club. As the annual weighing time for Lillian approached, the Liars' Club ran a competition in which the general public was invited to guess Lillian's weight for sixpence a try, with three prizes totalling £100. The donated money was for 'the assistance of sick and crippled children' in South Australia.

Giving applicants a couple of small clues, they were informed that the nearly seven-year-old Lillian was 6 feet 2 inches high (1.87 metres) at the last measuring and when fully grown could weigh between three and four tons. In fact in 1937 after the annual weighing she was 6 feet 4 inches high (1.93 metres) and, at 3220 lbs (1460 kilograms), weighed almost 1.5 tons. However, by 1942 when she had been weighed annually for six years, she developed such worrying nervousness (now

Lillian is weighed at the East End Market, circa 1937.
(Royal Zoological Society of South Australia Inc.)

known as 'panic attacks') at being led through the city streets, the annual weighing was abandoned. It was feared she might take fright and run amok, and she by then weighed almost two tons. As for liars' clubs, they existed in other cities for many years, but by the 1950s were known to be short of a few good liars.

A runaway elephant can cause chaos in the city streets. Perhaps zoo employees still remembered a circus elephant escaping in August 1924 from the old Exhibition Grounds off Victoria Drive where it had been tethered and running through Adelaide. Fortunately it happened at night, considering how much damage it could have done in the city traffic during the day. The elephant was first spotted soon after nightfall making its escape along Frome Road when there was little traffic about. A passerby looked in amazement when he made out the shadowy outline of an elephant, brandishing its trunk.

Its capture without pandemonium was possibly due to the good luck of the zoo director's son, Keith Minchin (future owner of the Snake Park – later named as the Koala Farm from 1927–1960), driving along Frome Road at that time. Although he couldn't avoid a collision with the elephant, another person may have been traumatised at the very thought. Though somewhat shaken, physically and mentally, when he saw the elephant sprawled on the road in front of his car he retained his presence of mind before leaping into action. He coaxed the elephant back on its feet by holding its ear and succeeded in calming it down. By this time, the zoo was informed that one of their elephants had escaped. Realising it must be a circus elephant at large, zoo attendants were sent off to look for it. They found Minchin with the beast and between them persuaded it to follow them to the zoo. After it was secured Minchin reminded the journalists of his knowledge of elephants and that 'of course, you can't catch hold of it by the trunk or the tail and drag it along'.

Downplaying the event, Minchin told journalists 'it was only a young fellow … and was apparently, only out for mischief. It couldn't do much harm. At present it is safely locked up, and is amusing itself by battering a bucket'.

As for much-loved Lillian, who never escaped to run amok or go walkabout on her own, she endeared herself to all who visited the zoo and her. On her seventh birthday in the Society's jubilee year (when it added

'Royal' to its name and sensibly dropped the word 'Acclimatization'), Lillian was celebrated. All children were invited to a party in May 1938 when a special cake was made for her from bran and hay and garnished with Brussels sprouts. The topping was of carrots serving as candles. Real birthday cake slices were distributed in bags with a penny to the first 1500 children who arrived at the zoo. For the children who missed out on cake, Lillian was on hand to give free rides, for after all it was just another working day for her.

As early as 1943, when Lillian was showing signs of being troubled by sore feet and internal pain, Director Vince Haggard stated he had on hand an Indian vet's book on elephants. He consulted it for possible remedies. In 1950, when Lillian's health problems were becoming a regular topic of conversation at the council meetings, there were plans completed beyond the zoo for another kind of elephant to take up residence in Adelaide.

This was Nellie the elephant, a robot the height of a full-size elephant (that could reach 8 feet 6inches high, that is 2.59 metres). Nellie couldn't bellow, trumpet or rumble like an elephant, but sounded just like a motor-car. Nellie was taken out of storage once a year on a drive rather than a walk around the city as part of John Martin's Christmas Pageant. Other than the annual pageant she made guest appearances at fetes and other worthy fundraising events. However, unlike a real elephant, for many years Nellie had a tardy track record because she regularly broke down. But after 1981 when Alby Hill and his son Ross from Norton Summit took responsibility for Nellie's smooth operation, she ran

Nellie the mechanical elephant in the John Martin's Christmas Pageant in 1952. (State Library of South Australia B 69658)

like clockwork (or so claimed a newspaper report of 1994).

After the Second World War, as the world returned to some normality, slightly eccentric inventors concentrated once more on dreaming up new-fangled inventions that included toys, robots and other mechanical gadgets. When the John Martin's store wanted to perk up their annual Christmas procession with something new and original added to its existing 30 floats, what better way than to acknowledge Lillian, the Adelaide Zoo's third elephant, with a motorised replica?

After the first robotic Nellie made her debut in England on 28 July 1950 on an Essex road, pictures were flashed around the world showing her carrying a group of children in her howdah. The owner of the John Martin's store, Edward Hayward, who lived at Carrick Hill House, immediately ordered one at £1500. And it cost that much again in freight charges.

When Mr Frank Stuart, a theatrical mask-maker from Thaxted, England, saw some donkeys on the beach in about 1948, he, inspired as only inventors can be, jumped into action to produce his first elephant. It was made as a tubular-steel skeleton from 9000 parts and was complete with a 10-HP car engine that drove the moving legs. Weighing about a ton, its steel innards were covered in a hide of specially toughened inch-thick paper.

As the elephant trundled along on eight wheels its head nodded while wagging its trunk. Like a real elephant it was steered from a seat behind its ears using protruding bars. The speed was controlled by a gear lever through the elephant's neck. Very cannily, the exhaust fumes escaped through its trunk.

After successful world publicity through newspapers, Mr Stuart immediately made a killing with six to be constructed for the Festival of Britain, three were ordered from South Africa, 14 from Canada, 10 from Australia, and £33,000 worth from the USA.

But at its outset, Adelaide authorities were quick to declare that their mechanical elephant was not a robot or a toy but a vehicle because it had eight wheels and a car engine that could reach a top speed of 27 miles per hour. When Mr H.B. Walker, the Registrar of South Australian Motor Vehicles, learnt that Nellie would motor through the city in John Martin's Christmas Pageant on her rubber-tyred wheels (that were actually small guide wheels in each bulky leg), he solemnly ruled on a £4 fee to register Nellie for six months. She also had to carry two number plates 'somewhere around her person' along with a fire extinguisher.

While the real elephants have long since gone from Adelaide, Nellie is still going strong after 64 years and is taken out of storage to be used in most of the annual Christmas pageants, now organised and sponsored by six South Australian credit unions.

Other inventors have had flashes of brilliance in odd places and situations, again associated with an elephant. In the *Advertiser* in October 1933 an article titled 'Queer jobs for women', informed of a Miss Flora Back who, as sister to the matron in charge of the Children's Convalescent Home at Mount Lofty, was an inventor of numerous devices for healing and comforting human ills – in other words she was a surgical appliances inventor. She said that when she was only eight years old in about 1899 the Adelaide Zoo elephant, Miss Siam, was in a playful mood when 'the brute ran away' with her in the howdah. She said the elephant twined its trunk around her in an attempt to put her down on the ground but 'I could not fall off, because I was strapped on, but the shock led to internal displacement'. She became obsessed with curing herself and invented a device to ease her condition. 'It was to cure myself that I invented the first of my appliances, and since then, with the help and encouragement of the medical profession, I have found in such inventions an absorbing occupation.' Amongst her other inventions were a special strap to save middle-age women from sagging chins, and appliances for curing flat feet and restraining what are politely known as 'outstanding' ears. Again she emphasised that her career took off entirely due to the runaway elephant at the Adelaide Zoo.

As to Lillian, in the years immediately following Nellie's arrival in Adelaide, the zoo council continued to be concerned about Lillian's health. In 1952 the issue of replacing Lillian with a new elephant was first raised by council member Dr Brummitt, who criticised her as a 'very poor specimen who was really unfit to pull the cart'. He suggested that she 'should be replaced'. At only 24 years of age, Lillian, in very poor health, was old before her time. Not only were the zoo staff aware of her failing health but a regular visitor, a young Trevor Klein, now a zoo volunteer, had noticed that she seemed to be getting meeker and weaker ... 'She started to look poorly and her whole physical demeanour changed, she became almost saggy as though her flesh was sagging on her skeleton.'

Diagnosed as suffering from rheumatism, she had developed stiff joints in her legs and had poor feet unresponsive to the treatment of sodium salicylate. In more recent years entire books have been dedicated to the specialised care of elephants' feet, so crucial are they to their good health. One particular publication on the subject was retailing at around US$2000 in 2011.

In 1954 any staff member who had charge of the

elephant was paid an extra shilling a day. And to help to pay for these expenses, the cost of elephant rides was to be raised from three pence to sixpence. However, due to Lillian's poor health she stopped giving cart rides anyway, able only to do 'light work' until she was officially retired later that year.

Shortly after, the Society's Exhibits Committee met to discuss the best method of disposing of Lillian. However, getting rid of an elephant is not the same as getting rid of an elephant shrew or even an unwanted kitten, and so advice was sought from both the Sydney and Melbourne zoos.

As author George Orwell recalled in his 1936 article 'Shooting an Elephant', he found out when a police officer in Moulmein, Burma, in the 1920s that it was difficult to kill an elephant. He had the hard decision of whether to shoot one that had run amok and trampled a 'peasant' to death. He recalled 'I did not want to shoot the elephant' but being in a crisis he had to do what the 'natives' expected of him and what would save face. After five shots the elephant still took more than half an hour to die, and he walked away after a botched job that earned him an unwanted job transfer.

And so it was reported of Lillian in the zoo annual report for 1956 that 'her condition deteriorated until there was no alternative but to do away with her'.

Almost any incident at the zoo was newsworthy and so killing an elephant was going to be nothing short of a sensation. Therefore, the disposal of Lillian called for careful 'hush hush' planning. Weeks after news of the safe arrival of Samorn, the zoo's fourth elephant, in Sydney in April 1956, the president, director and council consented to have Lillian shot on the Wednesday evening of 2 May around 5.45 pm, after the zoo had closed for the day.

The decision to shoot her was not included in the minutes circulated to the council members for the next meeting. At the meeting they were told and sworn to secrecy as it was important to the zoo that Lillian be destroyed and dismembered immediately afterwards as humanely as possible to avoid leaks to the press.

Sir John Keith Angas, a member of the zoo council and one-time president, was known as a good shot. His reputation dated back to 1922 when he shot a wounded lion that had attacked and begun mauling his Australian travelling companion, Captain George Outram, known among the locals as 'the white hunter'. The *Register* for 15 July 1922 recorded in an article 'Big game hunting' that the men were on a two-month safari in Kenya, 100 miles west of Mombasa, when the attack took place. It resulted in Outram's death a week later from his wounds. So because of events that had taken place more than 30 years before, Angas was chosen for the abhorrent task of shooting Lillian using a Rigby 350 Nitro Magnum (perhaps the same one he took with him on safari). The event was very much like a legal execution for several council members and three zookeepers were present. Armed with vital advice from two interstate zoos, Sir Keith aimed the gun from behind Lillian's left ear, and with the first shot she died instantly. Just to be sure, there were two more to avoid any possibility of her being only stunned.

Sir Keith Angas with safari trophy, 1922. (Photograph courtesy of Michel Angas and Emma Simmons)

Who can say what torrid emotions were going through the keepers' minds, for their association with Lillian was far from over. Because the relationship between an elephant and her keepers can be very close, the task that the three of them had yet to perform can only be described as onerous.

Killing an elephant was hard enough, but to dismember one was gruesome. Lillian's carcass was neither buried nor retained by the South Australian Museum or Medical School as were the previous two elephants. Not for her was the dignified option of being retained as a revered beast by having her carcass stuffed or displaying her skeleton. In the most undignified option, her final destination after dismemberment was the boiling-down works at Gepps Cross. And further, so that her remains would fit through a small entrance to a digester, she was cut into strips five feet long and 16 inches wide.

Under an electric light three keepers, Albert von Haacht, C.J. (Charlie) Adkins and J.D. Moyle, worked fast through the evening and completed the task by ten o'clock that night. The next morning they returned to prepare Lillian's remains for transport to the boiling-down works later that morning. Only when she had been dispatched did the zoo director inform the Chief Quarantine Officer of Lillian's demise before also letting the media know.

The three keepers were paid a bonus of £4 each for the unpleasant work they performed. Before Lillian was dismembered, Dr Watts, the honorary vet, said of her post mortem, her poor condition was due to 'chronic bursitis with haemorrhage'.

It seems that any child or adult lucky enough to have known more than one of the zoo's elephants in their lifetime can be adamant as to which one was their personal favourite. Perhaps it's to do with seeing one for the first time as a child, so the memory of Lillian is forever imprinted as a fond childhood memory. This is especially so for Trevor Klein, the zoo volunteer. When Lillian died in 1956 Trevor, who first saw her when he was a three-year-old in 1946, felt sorry for her because she'd spent quite a bit of her life in an enclosure that was too small for an elephant. Helping to keep his memory vibrant is a photo of him sitting under a Moreton Bay Fig tree on his first visit to the zoo. He grew up with Lillian and had an empathy with the peacefulness and docility she conveyed during her daily exhibition. And in his early years, his mother and grandmother, who lived in North Adelaide, were able to take him as often to the zoo as most small children go to playgrounds today. His grandmother would settle down to knit or crochet and he was allowed to wander freely, so long as it was in her field of vision.

Trevor recalled that 'going to the zoo started to build in me a great interest in the natural world. We didn't have books in our house ... As a small child I was very observant and consequently the zoo was like an encyclopaedia of animals virtually being opened before me, where everything moved off the page'.

Trevor said of Lillian's sad death: 'I've always known that everything has a life and life always ends in death and so I recognised it was a natural happening.' Therefore he was looking forward to the new elephant. But he remembered that Samorn was a totally different animal. Although only a young elephant, her personality was that of a healthy, frisky but 'difficult' beast. Despite having a deep empathy with zoo animals, he never felt closeness with Samorn in the same way he had with Lillian, but then there was to be a new generation of children who would see Samorn as their very first elephant and she would become their favourite.

4

White hunters and animal traders – Peter (Peif) Ryhiner

AS DETAILED IN THE PREVIOUS CHAPTER, in June 1954 the council members of the Society took up Dr Brummitt's suggestion of two years before and discussed the idea of a new elephant, to be the Society's fourth. This was prompted by Lillian's continuing poor health.

The zoo first contacted Taronga Park, which at that time had seven elephants, asking if they had a suitable one they wanted to sell, but there wasn't one on offer. It was a good time for discussion as a month before a slump in Indian elephant prices, falling from £225 to half that, had been reported in the British *Sunday Times*. The zoo also contacted the various circuses to see if they had a suitable elephant, but there were none on offer.

Because procuring an elephant could take some time, it was in November 1954 that the Adelaide Zoo first contacted Peter Ryhiner, a 'wild animal catcher', to supply them with an Indian elephant. Recommended by the director of Colombo Zoo as a 'reputable dealer', Ryhiner was one of the most famous hunters of the era supplying zoos and circuses throughout the world with the wild animals they needed.

Enigmatic, romantic, a charmer and adventurer, Peter (Peif) Ryhiner was born in 1920 in Basel, Switzerland, the son of a family doctor. With an interest in zoology in his teens, he went to Amsterdam over a weekend and bought from a sailor a parrot that he sold in Basel at a huge profit. This prompted him to consider a career trading in exotic animals but it was to be a few years before he could pursue it because the Second

Renowned Swiss animal trader Peter Ryhiner, circa 1950s. From The Wildest Game, *1959.*

World War disrupted the plans of many an ambitious young man like him. A year after the war was over the now 26-year-old Ryhiner decided to go hunting, declaring 'I made myself so impossible, the jungle was the only place left'. And so he set out 'to live my jungle life' in South America.

Arriving in the United States en route to South America, he was advised not to attempt the venture without an experienced partner. In Buenos Aires he met an old German, Lothar Behrend, one of the world's most famous, if not notorious, animal traders who, as the black sheep of a wealthy industrial Dresden family, proved to be the right kind of man from whom to learn the ropes of the trade. Ryhiner wrote that in spite of his abuse and open dishonesty, 'I knew I could learn more from Behrend in a week than from any other collector in a lifetime'.

Behrend was trading with the Adelaide Zoo from the 1930s to 1947 when in that year he sent stock that included a puma, ocelots, hairy armadillos, flamingos, and about 40 other birds in return for a consignment of Australian wildlife being sent to the Zurich Zoo. Further, because he had known Ron Minchin, the zoo director, very well before his death in 1940, he wrote to the Adelaide Zoo that 'I take the liberty to forward also to your esteemed institute a small collection of the most interesting and … easily-bred, grain eating small birds of Argentina … by way of memorial-gift to our unforgettable friend R.R. Minchin'.

So it was with this trader that Ryhiner threw in his lot, and only three weeks after their meeting they were off gathering animals for South American zoos. Ten years on, Ryhiner, who was then hunting and collecting on his own, had covered most of the jungles of the world and captured around 150 elephants and thousands of other animals.

Big game hunting, particularly 'on safari', was the domain for the rich who wanted to show off their wealth through their mounted animal trophies and ivory tusks. But by the 1950s the gun was giving way to the camera on safari; the tourist market had begun catering for those wishing to see large wild animals in their natural habitat. And zoos were still earnest about putting on the most exotic displays for their paying public and relied heavily on a 'white hunter' or an 'animal trader' to keep them well supplied. These characters often had tremendous celebrity status, and as Ryhiner emphasised, 'publicity is vitally important to an animal collector for unless zoos and big dealers have heard of you, they ignore your letters'.

Then, as now, celebrity often attracted the most beautiful and notable women. Mercia, a beautiful Eurasian woman from Singapore who was a clerk at the Adelphi Hotel then located in Coleman Street on the corner of North Bridge Road, married Ryhiner in 1950. She later moved on to be the 'close friend' of the author Graham Greene, and then the actor Rex Harrison's last wife.

During and just after the Second World War, zoos around the world could not easily obtain new species. As the aeroplane came into its own, large consignments of animals could reach anywhere in the world within 72 hours. This was a fabulously fruitful decade for organised animal traders familiar with shipping and air-freight. It was also a period in which there were fewer import and export restrictions to contend with.

From the mid 1950s, however, big game hunting for ivory, or other parts of an animal mounted as trophies, was rapidly becoming unpopular. And also no longer generally accepted was the capture of wild elephants for zoos and performance animals for circuses. The new buzzwords of conservation and the more alarming discussion of endangered species and extinction, began to affect the animal traders' business and more importantly, the relevance of zoos.

Peter Ryhiner, who was one of only about four postwar animal traders handling large shipments of animals in the 1950s, was having a difficult time adjusting to the declining popularity of his profession and the fact that the traders and hunters were rapidly losing their celebrity status. They were also facing ever-increasing restrictions as countries banned the export of their indigenous flora and fauna.

Peter Ryhiner's business was badly affected in 1950, when Thailand enforced restrictions on the export of its wildlife. In his publication *The Wildest Game*, he recalled being in Thailand and staying one night at the royal palace in Bangkok with his friend Theo Meier, a Swiss artist, who, usually domiciled in Bali, was also a guest of Prince Rangsit. Ryhiner was about to tranship a huge and very profitable consignment of captured animals that had taken him six months to collect for the Antwerp Zoo when Prince Rangsit's father, the Thai Prince Regent, over cognac and the finest Havana cigars after dinner in the palace's impressive dining room, declared to him casually, 'Oh, by the way, I just signed a bill that will affect you. All shipments of wild animals from Siam are forbidden.'

In one stroke of the pen six months' work of collecting animals was for nothing and Ryhiner was unable to even pay for his passage out of the country. But as Ryhiner had learnt from being in the trade, one needed a network of trusted friends or contacts, such as handy senior conservation officers in specific places who could help out a mate. In this case he knew someone in Thailand who could help him out of the potential debt crisis caused by the Prince Regent's sudden edict by putting him in touch with a senior conservation officer of forests in Assam. That person would help him fulfill his next animal order for the Basel Zoo in Switzerland. In the meantime Ryhiner was able to sell many animals of his Thai consignment to local traders.

As countries around the world legislated to protect against trafficking of their own wildlife, it is a sorry fact that some wildlife officials and park rangers became important links in major and successful smuggling rackets, particularly in Australia.

Four years after Ryhiner's encounter with the Prince Regent, in 1954 the Exhibits Committee of the Adelaide Zoo met to discuss the best method of disposing of their ailing elephant Lillian and replacing her with a young one from India supplied by Peter Ryhiner. Adelaide had been recovering from its only serious earthquake to date, which caused minor damage to many of its city buildings – including the zoo director's residence that lost its chimney when if fell off and smashed into the iron roof cutting electric wires in the process.

Far away from Adelaide, when animal traders and white hunters were still looked at in some awe, Ryhiner wrote to the zoo director that he was able to secure an Indian elephant: 'I am in the position to supply you a female elephant, its health in good condition, between 5' & 6' in height, that means between 6 & 8 years of age … The price can be quoted Rs 7000 [rupees] delivered to Colombo Harbour.' Indeed, at this time, Ryhiner was shipping large numbers of juvenile elephants out of India and Ceylon to America to fulfil the heavy demands for young female elephants there. It was within this period that Adelaide made a request for a 'well grown female, approximately 6–7 years and not to exceed £100 per foot'.

On these terms the Adelaide Zoo continued with the contract with a slight amendment by Peter Ryhiner that the elephant would be shipped direct from Cochin to Australia. But although this deal went ahead, perhaps it had been overlooked that only a month earlier the director had casually mentioned to the Minister of Lands, Mr C.S. Hincks, that the Adelaide Zoo was looking for a replacement elephant for Lillian. The Minister had relayed the comment when he was at an annual meeting with the Thai Legation in Canberra.

Nothing more happened until the Minister of Lands received a letter out of the blue in February 1955 from the Thai Legation, announcing 'our authorities will be able to provide an elephant'.

At this stage it was not specified whether the

elephant was to be bought, a gift, or an exchange. The zoo director, Vince Haggard, phone the Legation. The Thai official told Haggard that he was going to suggest to his government that the elephant should be a gift but when he heard a rumour that the Adelaide Zoo had ordered an elephant from elsewhere, he was no longer sure whether they wanted a second elephant at all and so had not contacted his government in Bangkok to secure one.

During this period of negotiation the media in Bangkok had learned that Adelaide was looking for an elephant. Lavington Bonython, the proprietor of the Adelaide *Advertiser* newspaper, rang the zoo director on the night of 18 April 1955, informing him that he'd received a message from Reuters Bangkok that 'a request from Adelaide Zoo for a Thai elephant had reached cabinet level here. They were under the impression that an elephant had died in Adelaide. The Council of Ministers agreed to send the elephant to Adelaide to replace the one that died here …' Mention of a Thai elephant had already appeared in the *Advertiser* on 12 April.

After the director's next phone call to the Thai embassy he declared, 'I think they would appreciate a donation of kangaroos and birds if we accept an elephant.' Before the introduction of draconian national laws in December 1959 that all but stopped the native trade dead in its tracks, it was a very easy matter to donate, sell or exchange any number of Australian native flora and fauna without recrimination, red tape aside. This is how zoos were able to obtain new species so easily for their collections or find new homes for surplus indigenous species.

Although the Act of 1959 was prohibitive, London's Regent Park Zoo had decided as early as 1950 not to have anything to do with any Australian zoo because of the excessive red tape, with lengthy forms filled in triplicate, statutory declarations and other quarantine regulations imposed by the Commonwealth government. All of which was described as 'ridiculous' by the London Zoo's superintendent.

The Adelaide Zoo director wanted to accept the Thai elephant, provided it was 'partly trained prior to shipment'. The zoo council, carried away with its elephant plans for almost a year, had not realised that two elephants were about to arrive when there was not enough room for them both. And Lillian was still very much alive.

It was only when the 'gift elephant' from Thailand, who was given the name Samorn – 'beautiful and beloved' – arrived in Sydney a year later and the zoo had to pay for the exorbitant shipping bill, the telegrams, food and so on, that it became evident that receiving her actually cost the zoo more than £550 – quite apart from the consignment of Australian animals they made in exchange. In all, it was a costly deal and not at all within the true meaning of a gift.

The zoo then realised that Peter Ryhiner's Indian elephant was also arriving shortly and perhaps after all they no longer wanted two elephants. But the transaction had gone too far, and because there was no way out of the deal they decided to accept the second elephant anyway and try to sell it on arrival to Wirth's or Bullens Circus.

But as time wore on, the interested circuses were only prepared to buy the Indian elephant provided they could inspect it when it arrived. This was a risk the zoo simply couldn't afford to take. If it was rejected, they would be left with an extra elephant they could not really afford or find space for.

So they reluctantly cancelled the contract with Peter Ryhiner and called in their lawyers to sort out what could be a costly fee for defaulting on it. By careful negotiation, the zoo got away with having to pay only £210 to Ryhiner, which also included the cost of the fodder for the elephant he was caring for in India while waiting for a ship to transport her. He could have been meaner, but he wasn't.

It is not often asked where a captive elephant has actually come from. Was it rescued from a circus or did it come from another zoo? Nearly every elephant in captivity until fairly recent times was born in the wild. How it ended up in a zoo or circus was rarely questioned. If Samorn was captured from the wild as a six-year-old, could she have been rescued when her mother was killed or was she part of an elephant hunt, once so frequent in Thailand?

5

A Siamese elephant hunt of 1939

ALTHOUGH SAMORN WAS MOST LIKELY captured from the wild, this was not as part of the once traditional royal elephant hunts that took place until about the end of the 1930s. Until then, for at least 400 years, elephant hunting in Siam was a royal event conducted by the court for the kings, their family and friends. For everyone else, although elephant hunting was prohibited by law, special permission was granted to groups of hunters wanting to capture wild elephants to train them for labour, such as for the heavy work in the teak forests.

Samorn may have come from Lopburi or Ayutthaya, two famous elephant hunting localities about 100 miles north of Bangkok (and the place of great floods in October 2011). The once annual royal elephant hunts, now organised by the Thai military, were rarely held but when they were, it was to raise funds for the equipment of a new cadets' school, or for other such items.

It was in this part of Siam where the kraal fashion of capturing elephants was traditionally carried out.

One of the last major annual elephant hunts, according to newspaper accounts, was conducted in late 1939 at Lopburi just as the Second World War began in Europe. The upkeep of the elephant kraal there, paid for by the Thai government, was now overseen by the military when it moved its headquarters there. However, after the war the King of Thailand arranged one more royal elephant catching ceremony in January 1962 to honour a visit by the King and Queen of Denmark.

Elephant kraal in Ayutthaya, Thailand, in 1864 from Henri Muhout, Travels in the Central Parts of Indo-China.

Newspaper accounts for the 1939 hunt stated that the 'whole population of Siam' was stirred with excitement over the elephant hunt that took place at Lopburi. More than 100 tame elephants with their mahouts set out to bring back as many wild ones as they could herd into the massive kraal. Near where the hunt was carried out in the depth of a great forest was a large open 150 by 100-metre kraal enclosure, conveniently located in a valley at the foot of two hills. The natural features were used as galleries for the thousands of spectators who made the journey to watch the exciting spectacle of wild elephants being herded.

The kraal, which was constructed entirely of massive teak timbers, was fenced with huge strong wooden spikes pointing inwards, while inside was a ditch encircling the ground. At the entrance to the kraal was a huge timber gateway through which the wild elephants were herded. A moveable beam of the gate was attached to another beam on the top so, while wild elephants could enter, there was no escape. When the hunt was over the tame elephants with their mahouts left by another exit. It was also through this gate that the unsuitable wild elephants were later set free.

The hunt was a magnificent sight as 150 tame and wild elephants lumbered one by one through the palisade and the big gate into the kraal. Before the elephants were herded they funnelled through a strong palisade erected in the form of a 'V' so the passage became progressively narrower as they approached the gateway. All too often it was at the palisade that the wild elephants first sensed the danger of being captured and tried to escape.

The mahouts did their best to control about 50 wild elephants led by a huge tusker. The tame ones were just behind, guiding the wild ones in the right direction, but when they approached the palisade there was confusion when the wild leader, sensing his awful fate, refused to enter and dashed back into the herd, trumpeting and bellowing. The mahouts rushed to guide him back clapping their hands and shouting at him until he turned and headed once more into the kraal where he ran about wildly, much upset at his plight.

It had taken about two hours to herd them to the arena. None of the elephants were hurt but they had become distressed once they knew there was no escape. The elephant hunters, drawing on centuries of practice of dealing with newly captured and distressed wild elephants, had a remarkable way of calming them down. They used the salve of music.

A newspaper account reported that 'queer music started, lullabies were sung and played by a group of selected musicians from northern Siam [most likely Surin], to quieten the excited animals'. Only when the wild elephants were calmed and soothed did the musicians stop playing and the tame elephants leave the arena.

Causing a mild sensation during this roundup was a month-old baby elephant that had naturally followed his mother through the gate. The next day, the best specimens were selected and retained to train for labouring work. When it came for the unsuitable ones to be driven back to the forest, tame elephants accompanied them to make sure crops and properties in the rice fields were not damaged on the way.

In times past the elephant hunt took place over several days, and used the last day much like a circus entertainment, except that the wild elephants were usually taunted and teased to put on a show for the spectators, by enticing them into rages of bellowing and trumpeting. But that was all a long time ago.

Currently, tourists visiting Thailand can visit elephant sanctuaries in places like Ayutthaya, Lopburi and the district of Chiang Mai. Tourists prepared to pay generous donations can look after the elephants, ride and bathe with them for a day or over several days. One can also see wild elephants in such places as the World Heritage National Park of Koh Soi. Although elephants are a national treasure their plight is nothing short of a national tragedy. Their numbers in the wild have now plummeted below 2000, while that many again are domesticated and mostly drawn into the tourist trade. Alarmingly, increasing numbers are poached from Laos, Cambodia and Burma for 'domestication' to satisfy the Thai tourist trade. Fortunately, there are several charities doing their best to buy as much forest area as can be procured to accommodate the increasing numbers of old, infirm and homeless elephants.

Until 2013, apart from knowing that Samorn came from Thailand as a gift from the government, little was known in South Australia of her origins. There is nothing of detail in the Adelaide Zoo archival records, nor in the newspapers or from any other source, so I thought I would try and gather what information I could from the Thai Embassy in Canberra, believing it was a good starting point. But after a year of no replies to my letters in the quest for information, I gave up, especially after the Thai Consulate in Adelaide thought my chances of gathering useful information from Canberra might not be good.

This didn't help the scepticism I developed about finding material beyond South Australia. This changed when my partner, Robert Martin, read my draft manuscript and helpfully advised 'your manuscript is not complete until you go to Thailand yourself'. Having been told the bleeding obvious, I reluctantly agreed with him. However, where would I possibly start looking within Thailand if I could find nothing from their embassy in Canberra? What were the chances of finding a file languishing there in an archival repository? But I began to google, as we all do these days, and discovered the National Archives of Thailand. When I couldn't find a way through its email address, I wrote them a long detailed letter with my curious but specific enquiry.

No one was more surprised than I when, within 10 days of my posting the letter to Bangkok, the National Archives contacted me by email with the wonderful news that they held a 71-page Foreign Affairs file about Samorn.

Imagine my excitement at finding out about something I had thought could not possibly exist? However, my joy was tempered somewhat by discovering that at least half of it was written in Thai, when my Thai language is non-existent. But when I asked the Bangkok Archives to recommend a translator, they did so.

I knew this wonderful information existed, but the problem of accessing the Thai National Archives was far from straightforward. To even cross the threshold, one has to fill in a mass of paperwork from the National Research Council of Thailand (NRCT), giving research outline, referees, research grant costs, the project's relevance to Thailand, and so on and so forth – plus three photographs of oneself.

Not supported by a research grant, nor having an employer or attached to a university, I asked and received a much-needed sponsorship letter from the incoming CEO of the Adelaide Zoo, Ms Elaine Bensted; a letter, I believe, crucial to my obtaining permission to see the awaiting archival file.

Would a visit to the archives in Bangkok reveal where Samorn came from and whether she was born in captivity? Or whether she was an orphan or captured in the wild? I hoped the information I derived from the file would determine which part of Thailand we would then visit after Bangkok.

My partner would accompany me to Thailand and we would take the opportunity to make it a holiday as well as a research trip. In October when I booked flights to Thailand for late January 2013, I thought four months to gain permission to use the Thai National Archives in Bangkok would be ample time. We made it – but only just.

In the meantime the Thai bureaucracy partially thawed, for within a short period of time the Thai translator, the clerk at NRCT, the archivist at the National Archives, and I were all contacting each other by email on affectionate first name terms, which really surprised me. Further, I came to know about my translator, Patsri Tippayaprapai, of her husband and little boy and all about her time in Australia undertaking a university degree in Queensland. However, when it was almost Christmas and already nine weeks since I first contacted the National Archives and the National Research Council, I panicked when I realised how little time was left before I flew to Bangkok. I still had not received any official permission to use the Archives. I sent the NRCT an email to convey my concerns and was quickly informed that biting into the administration time were several public holidays, the King's birthday, Constitution Day and Christmas. I really despaired when I learned my application had passed successfully through their department, but had now been passed on to the Department of Fine Arts. This was an added blow for I hadn't known of this department's involvement. With only three weeks to go, I wondered whether I'd receive permission for research before I arrived in Bangkok. On 2 January there was a reassuring email stating I would be granted permission before 25 January, my flight date. But there was a sting in the tail when I was informed that I must pay for a non-immigrant visa, and pay within seven days of arrival a fee of 10,000 Baht ($300) to the NRCT, a form of guarantee of the lodgment of my book about Samorn with them on its publication. Some were quick to suggest this was nothing short of a bribe.

When I made a visit to the Thai Consulate in Adelaide to obtain a non-immigrant visa, the Royal Thai Consul, Jaye Walton, was on one of her many official visits to Thailand and her assistant David Wright, who looked after my application, told me wonderful tales of Thailand. When I returned to pick up my visa a few days later I was lucky enough to meet Jaye Walton, who had returned from Thailand. I reminded her that I had spoken with her by telephone the year before about my unsuccessful dealings with the Canberra office.

Jaye Walton was the one-time hostess of the TV show 'Touch of Elegance', shown for 12 years from 1968 on Channel 10 and Consul General for more than 30 years. She was the one to evacuate her studio when the Adelaide Zoo director attempted to take a boa constrictor there in 1980.

Jaye grew interested in my venture when I revealed I needed a visa to undertake research about Samorn. Because Samorn was from Thailand, Jaye was sometimes present at zoo events that involved her. So she knew Samorn very well. Formalities gave way to offers of kindness in so many ways, reassuring for one who had never been to Thailand. I cherished the small but powerful card embossed with the Royal Thai Government coat of arms Jaye gave me to help me out of difficult situations. I would show it to police or officials should I need to. Fortunately, this didn't eventuate. The Most Exalted Order of the White Elephant is the highest award bestowed by the Thai monarch on individuals for service to the nation and Jaye Walton has earned this award – twice.

By the time we left Adelaide, all government requirements were fulfilled. We at last arrived in Bangkok and had three days to visit its many attractions before meeting my translator, Patsri Tippayaprapai, whose nickname is 'Mameaw'.

She arrived at our hotel in Banglamphu on a Tuesday morning in her mother's car and drove us to the National Archives only five minutes away. Here I met the helpful archivists who showed me the file on Samorn. While it was being photocopied we visited the Ananta

Samakhom Throne Hall where the Thai monarchy hold many formal events.

After picking up the copy of the government file, we retreated to Mameaw's office conveniently located alongside the National Archives, and set to analysing each page. It didn't take long to realise that the crucial information I was looking for – Samorn's exact provenance – was not discussed and an element of detective work would be needed to come up with answers.

In the meantime, the file revealed other useful information. When South Australia requested an elephant, in 1955 Luang Chamnan Auksorn, the Thai Secretary General of the Administrative Council of Ministers, wrote to the Permanent Secretary of the Minister of Foreign Affairs that the request be passed on to the Ministry of the Interior. It replied that the Zoological Park Association (Bangkok's Dusit Zoo) would look for 'a friendly elephant which is around 6 years old'.

While the records in Bangkok did not reveal from where the young elephant was obtained, according to the Bangkok Dusit Zoo they had nevertheless obtained her and had 'already prepared the elephant' in 1955. As the zoo was informed that the elephant 'needs to serve children', she was then trained either to let them ride her or 'use her to drag the rickshaw around the zoo'.

Bangkok's Dusit Zoo was given a budget of about 26,000 Baht (about £500) by the Thai Ministry of the Interior. This sum covered the cost of searching for an elephant, preparing her for the voyage and several months of training as well as undergoing the necessary health checks to qualify for importation to Australia.

During her stay in Dusit Zoo she was under the care of Prince Chananwatra Tevakul, the Honorary Veterinarian and a committee member of the Zoological Park Association.

As Samorn was accommodated at the Dusit Zoo for some months before her voyage to Australia, when Mameaw suggested we make a visit, I was eager to go. When you first enter the zoo you see the 12-acre (4.5-hectare) lake within its tropical environment; indeed a beautiful place, right in the heart of Bangkok. I walked around most of it and was impressed with its setting – but that was about all. Once I saw seals performing to large crowds and two elephants chained by the leg and rocking and weaving about, confined in an area no larger than six metres square after performing their circus tricks to the appreciative public, I wanted to leave.

The 47-acre (approx. 19-hectare) Dusit Zoo, commonly called 'Khao Din', opened to the public in 1938 and was managed by the Bangkok City Municipality until 1954 when the property was transferred to the state Zoological Park Organisation.

Before then, however, the zoo site was originally part of King Chulalongkorn's private estate, known as Suan Dusit or Celestial Gardens (Khao Din Wa Na). He was inspired to build this 76-hectare (almost 188-acre) estate after visiting Europe in 1897 where he saw mansions and palaces amid landscaped gardens.

This new estate included 13 royal residences and several throne halls. His new timber residence, the Vimanmek Mansion, was dismantled and transferred to this site, and was the 'largest teak mansion in the

The Ayutthaya elephant kraal in 2013, photograph taken by the author.

world'. This was his primary but unofficial residence for five years from 1901. His new estate was located four kilometres from his official residence, the Grand Palace by the Chao Phraya River. He later stocked his gardens with the rusa deer (*Rusa timorensis*) that he brought back from his travels to Java.

But after he died in 1910, the gardens were neglected; first by his son, King Vajiravudh (Rama VI), who died in 1925 at 45, and then his son's younger brother who became King Prajadhipok (Rama VII). Until he abdicated in 1935, the gardens remained forgotten. A council of regents reigned on behalf of Ananda Mahidol, the boy who was only nine when he became the next monarch. Prime Minister Field Marshall Jomphol Por Phibulsongkram requested in 1938 that the regency transfer 47 acres of the gardens to the Bangkok Municipality. Together with the descendants of the original Javanese deer herd, this portion of the gardens became Bangkok's public zoo.

Not having found all the answers I was looking for in Bangkok, I thought we should travel first to Ayutthaya, the former capital of Thailand and one of the places where royal elephant hunts took place until 1906. For most of its history, the elephants were captured and trained by the Thai army to fight the monarch's battles, such as against the Burmese. After the Thai monarch moved the capital to Bangkok in 1781, Ayutthaya remained important for the elephant hunts.

The town boasts royal summer palaces, numerous temples and a 16th-century elephant kraal, the only kraal left in Thailand. To celebrate the present King of Thailand's 80th birthday in 2007, the site was renovated. It was originally created from nearly 1000 ageing round teak stakes half a metre in diameter driven deeply into the ground 30 centimetres apart and almost three metres high. This created a massive quadrangular stockade, capable of accommodating well over 100 elephants at a time. At the head of the quadrangle is the elegant Ko Jawet Maha Prasat Pavilion of traditional Thai architecture where the King and his court viewed the proceedings.

Many of the current tourist activities in this town focus on elephants. Indeed it is difficult to avoid them for they are found in several major tourist locations where they can be seen or ridden.

Making these activities possible is one man, Laithogrien Meepan, affectionately known as 'Pi-Om', a zoologist, who on the spur of the moment in the mid

1990s gave his daughter an elephant as a present. After realising what he had done, for it cost at least 600 baht (AU$30) a day to feed one, he began to investigate and understand the plight of Thailand's elephants generally. By 2000 he had seven elephants, and on 1 April 2005 established the Phra Kochabaan Foundation to fund his sanctuary The Ayutthaya Elephant Palace and Kraal Village. He now has well over 100 elephants. He is committed to providing a safe haven by buying and retiring old working ones along with those that are sick, abused, injured and even those that have killed humans. He employs and houses the mahouts who look after them.

His foundation grows food to feed his elephants as well as buying land for them to roam in. The foundation breeds elephants as well. Through his care, his elephants have had 53 successful births since 2000, however, he states that Asian elephants are dying faster than they are breeding. It is his life ambition to restore the Asian elephant to its rightful place in Thai society, as a much loved and respected symbol of Thailand's history.

6

The arrival of Samorn in Adelaide

IN FEBRUARY 1956, the Adelaide Zoo director received a letter from the Thai Ambassador in Canberra acknowledging the safe arrival in Singapore of 'the gift elephant Samorn'. She had been trans-shipped from Bangkok to Singapore into the care of Herbert H. de Souza of the Mayfield Kennels where she stayed until there was a ship to carry her to Australia. The ship *Idomeneus* set sail on 4 April with Samorn and arrived in Sydney on 20 April. She was taken to Taronga Park for quarantine purposes.

Several council members of the Adelaide Zoological Society went to see the new elephant at Taronga Park while visiting Sydney on business, and all agreed she 'appeared to be a very fine specimen'.

After Lillian was put down and her carcass disposed of, the elephant house was repaired, repainted, and installed with a wood-blocked floor in preparation for Samorn's arrival. The zoo had experienced many problems with the right type of flooring for an elephant.

Sand and soil were reduced to a quagmire in the winter, concrete damaged the feet, while timber floors, easily worn, had to be frequently replaced.

When the weather turned cold across southern Australia, the council members of the Society all agreed with Sir Edward (Ed) Hallstrom, the Taronga Park president, that 'the animal might suffer serious consequences if the trip [to Adelaide] was undertaken in the present exceptionally cold weather', and so Samorn stayed on in Sydney to have the benefit of the warm quarters provided for her. In avoiding the danger of transfer during winter weather, Samorn benefitted from more training at a 'critical stage in [her] career' although it cost the Adelaide Zoo a further £8 in insurance for her to stay in Sydney until the weather was milder.

Samorn may have thought Taronga Park was her new home and have been pretty surprised that after more than six months there, she was marched up a ramp into a semi-trailer lorry to be taken to another zoo a thousand miles away in Adelaide.

How was she to know this was going to happen? She'd settled down quite easily at Sydney Zoo because she had been fussed over by three matronly Asian elephants in the next yard named Sarina, Ranee and Jill, who all wanted to mother her. She didn't mind that, but they constantly squabbled with each other as each tried to grab her with their trunks to be next to her, even though a fence divided them. While only one had been a proper mother herself, they were overly protective and bossy about how they thought Samorn should behave and told each other how they should behave towards her.

Now, as preparations were made to board the semi-trailer, Keeper David 'Chick' Cody stroked Samorn's trunk regularly telling her that everything was going to be OK and that he was going to be in the front of the lorry with the driver over the next four days.

Perhaps he may have reassured her by blowing into her trunk and saying, 'Now don't you worry, my little Sheila, I'll be coming with you and we'll stop on the way to give you a break and a regular hosing down.'

Of course, Samorn had no idea what he meant by a break. She travelled in a trailer on the back of a lorry carrying a consignment of Sir Edward Hallstrom's refrigerators being delivered to the firm of Godfrey's in Adelaide. Keeper Cody led her to the lorry waiting at the entrance to the zoo and up the ramp into an eight-foot-square crate covered with a canvas tarpaulin. He said the tarpaulin was to make sure she didn't grab with her trunk the tram cables or the telephone wires on the journey. 'Now you behave yourself, Samorn,' Cody may have said and in reply perhaps Samorn brushed his cheek with her trunk. Sir Ed may have been at the zoo to see Samorn led onto his lorry bound for Adelaide. But then he was there for nearly every zoo event, no matter how minor.

The use of a fridge lorry was made possible by Sir Ed because he made his fortune through manufacturing fridges, particularly his model Silent Knight. And because of his own private zoo, it is alleged he made a financial killing through his illegal dealing in Australian native wildlife.

As Taronga Park's president but also its self-proclaimed director, vet, conveyor of all zoo wisdom and knowledge, he could, would and did promise the world when it came to donation, exchange or trade in whatever animal or bird he wanted – or that he thought Taronga wanted – when it suited him. Nothing was too much for him and most (if not all) of the Australian zoos turned to him first if they had a problem with an animal or their own zoo's administration. Sir Ed, who always had the answers and whose word was law, was called 'the Chief'.

And why would the Adelaide Zoo not seek his advice? He had after all been an Adelaide Zoo member since before the war following his generous financial donation to enable the zoo to construct much-needed bird cages. For this he was made a life member; in October 1947 the *Advertiser* referred to him as 'the godfather of the Adelaide Zoo'.

Whatever his failings, he wanted to be seen in the forefront of animal protection and conservation in all its forms, which included searching for recently declared extinct animals. In 1952 Sir Ed was organising an important expedition in search of one of the most fabled extinct animals in the world, the Tasmanian tiger (thylacine). It had not been seen officially for 20 years since the last one died in captivity in the Hobart Zoo, but regular sightings in the wild continued. He chose his son to lead the expedition. Steven Smith, a Tasmanian wildlife officer, claimed in 1980 that 'it has not been proven that the thylacine is extinct, nor that it survives, but the continued occurrence of sightings of thylacine-like animals provides some cause for hope that the species ... still exists'. This he concluded when he wrote an official obituary of the tiger in that year for the Tasmanian Parks and Wildlife Service. But remember Penny Olsen's words about proving extinction.

To Tasmania's great shame, only in the 1930s when the tiger had apparently been hunted to extinction did the bounty for their heads cease. And by then, when it was too late, they were made a protected species, and plans made to establish a sanctuary for them in the densely timbered and uninhabited Raglan Range east of Queenstown. The demise of the unusual striped marsupial has generated growing numbers of thylacine seekers who document their every sighting in regular bulletins. Growing numbers of writings are also guaranteed to keep the legend alive, such as Louis Nowra's *Into that Forest*, Julia Leigh's thriller *The Hunter*, Peter Goldsworthy's *Honk if you are Jesus*, and Col Bailey's books about thylacines, including his bestseller *Shadow of the Thylacine.*

Although Tasmania attracts much criticism through its loss of the thylacine, the mainland of Australia has a far worse track record, for by 1964 'as much as 40% of Australia's mammals are so rarely seen as to be considered extinct, and many classes of parrots and finches have virtually died out', according to a source for *The Bulletin* on 4 July of that year.

Sir Ed showed the right sentiment about conservation at this time. But his reputation for trading in Australian wildlife was known far and wide. Even one of the world's renowned animal traders, Peter Ryhiner, had something to say of his activities in his book *The Wildest Game* when he remarked that in the years before

Sir Edward Hallstrom with a black cockatoo at Taronga Park, 12 November 1964. (Fairfax Syndication)

1950 Australia had been a closed market to animal collectors as the government rigorously protected its fauna by daunting red tape. Sir Edward Hallstrom was an ardent parrot enthusiast and before the Second World War Lothar Behrend (Ryhiner's former partner) had promised to send him an almost unknown species of the Conure parrot from the Andes, as well as a large shipment of animals from South America, in exchange for Australian animals.

True to his word, Behrend sent the shipment containing the parrot and other animals, and Hallstrom's shipment arrived soon after in exchange. Described as a 'breathtaking wonder', the Australian consignment worth, even then, about US$120,000 (a small fortune by the standards of the day), included kangaroos, wombats, wallabies, Tasmanian Devils, echidnas, hundreds of parrots and other rare birds of all descriptions, and reptiles.

The shipment was sent via San Domingo, then briefly known as Cuidad Trujillo, in the Dominican Republic, where a new zoo was being established. While some of the animals and birds stayed at the zoo, much of the consignment was bound for the United States. It was well known that consignments such as this never went directly to the countries ordering the animals, but went via others as part of an elaborate laundering process to cover tracks and avoid too many questions being asked.

In this case, some animals, especially parrots, could not be imported directly into America because of the disease psittacosis. However, the small Caribbean republic, then being 'sweet-talked' by the United States not to turn communist, was seen as a back door into the US and Ryhiner wrote that 'as part of the American programme to placate Dictator Trujillo animals can be imported ... with very little trouble just by unloading the entire cargo there for a few weeks so as to establish their origin as being the Dominican Republic'.

He was a good bloke, Sir Ed, energetic and generous. He had time to build up his own private menagerie that included the rarest of Australian cockatoos and parrots – and he was only too happy to trade, exchange, buy and sell from Taronga Park and his own collection. All may have continued smoothly for Sir Ed, but the new and draconian federal *Flora and Fauna Act of 1959* tripped him up, mainly because it was too difficult to change the habits of a lifetime.

Sir Edward Hallstrom received honours, accolades and generous sentiments in regard to his generosity and philanthropy (he was president of Taronga Park and was the zoo's greatest benefactor – he'd donated more than $500,000 in cash and as many birds and animals of equal value). But in his older years, when he could have basked in glory, he was much criticised and sometimes referred to as Mr Concrete for the amount of it used at Taronga Park. In the *Australian Dictionary of Biography* it is stated that 'in 1966 he was also under covert surveillance for illegal trafficking in fauna. Four years later, 35 people were convicted of that offence and it was thought he may have pulled strings to have his involvement concealed'. It was

Samorn is coaxed into the crate on the lorry at Taronga Zoo in November 1956. (Photographer Douglass Baglin; courtesy of Yvonne Austin)

only after his death in 1970 that he was openly denounced as an Australian wildlife smuggler. Unfortunately, such condemnation also had unfair repercussions, for several Taronga zookeepers were, for years after, tainted simply because of their association with Sir Ed.

But on this day in November 1956, Sir Edward may have been at the zoo to farewell Samorn. There was only a foot of space between the top of her head and the tarpaulin in the lorry for the ride to Adelaide. Dave Cody said Samorn was delightful when they stopped about six times daily to cool her off. When they left Sydney at 8 pm Samorn was a unsettled by the roar of trams and trains as they crossed Sydney Harbour Bridge, but she soon settled and was extremely good tempered. But then any elephant would have felt comfortable in the company of Cody. Described as having the posture of a confident cricket player, he was slim and erect with blue eyes and a handsome tan. Awarded the Medal of the Order of Australia (OAM) in 1990 for 'service to animal welfare, particularly at Taronga Park Zoo' and known as 'Mr Cool' or the 'Crazy Zookeeper', he conveyed a calmness and a liking for long drags from cigarettes.

Fortunately, Samorn could see the country behind the lorry. She may have noticed that the landscape was quite different from the forests of Thailand. While this place was flat, dry and desolate, travelling at speed created a cool draft, which was appreciated for the late November weather was already hot.

Every now and again when the driver stopped at some roadhouse, Keeper Cody would see to Samorn, filling the water trough and giving her treats of apples and bananas, and then leading her out to hose her down. Wherever this happened, a crowd of adults and children gathered to see her.

When the semi-trailer arrived in Murray Bridge in South Australia about teatime, there was a crowd of children waiting to greet her with apples, carrots and other fruity tidbits and they were captured in a media photo with Samorn.

During her journey Samorn had chomped through such foodstuffs as the tops of two bales of oaten hay, 50 pounds of carrots, six tins of condensed milk, several pounds of sugar (including the paper bag), 30 pounds of sweet potatoes, and 25 pounds of potatoes and bran mash.

They spent their last night of travel in Murray Bridge. Keeper Cody told Samorn this was so they could the next morning make 'an entrance' into Adelaide like the Queen of Sheba. Probably he told her there would be hundreds of children at the zoo to see her arrive.

There was great excitement leading up to the arrival of the zoo's new elephant, but her journey from Sydney had been delayed because of Adelaide's civic by-law banning semi-trailer vehicles of the kind that Samorn was travelling on from coming into the CBD after 8 am. Before the journey took place Sir Ed applied to the Adelaide City Council for special dispensation to have the elephant driven to her new home through the centre of the city on the back of one of his refrigerator lorries. Recognising the importance of the zoo's latest attraction of a new elephant for children to love, the council graciously granted a special permit for the truck to travel through city streets so people would have a chance to glimpse her passing by.

But priorities being priorities and business being business, the truck was also driving to Adelaide to deliver a consignment of Sir Ed's refrigerators to the electrical retailers, Godfrey's. The 21 refrigerators were off-loaded first in Keswick early in the morning, and then about 10 o'clock the lorry drove to the zoo by way of King William Street where it was hoped the route would be lined with excited children.

7

Early days in Adelaide

AFTER A THOUSAND-MILE ROAD JOURNEY over four days Samorn made her entrance into Adelaide a day before the beginning of summer on 30 November 1956. Following an extended tour of the city the semi-trailer pulled up beside a large earthen mound at the zoo. There Samorn was unshackled from the chains around her ankles that were secured to the corner of the lorry to prevent her from upsetting it while moving. She then walked easily from the trailer into her new home.

Welcoming this ton of charm were scores of children. Keeper Dave Cody, who'd spent six months training her at Taronga Park, said Samorn was 'the friendliest and smartest' of 20 elephants he had known. 'She was a good girl. She's got a sweet nature ... but she trumpets louder than most.' As he said this Samorn poked her trunk through jarrah bars of her crate trumpeting loudly for all to hear.

According to the newspapers Samorn, a teapot-shaped

'Chick' Cody of Taronga Park, Sydney, travelled with Samorn to Adelaide. Sunday Mail, *27 May 1956.*
(News Ltd – Newspix)

lady from Thailand, puckered her lips and wheezed as though to say, 'It was a cool trip, man, real cool.' Spending four days on the road on the back of a semi-trailer with a load of refrigerators had apparently had a psychological effect for Samorn arrived as cool as a cucumber and in good humour.

Supposed to be waiting for her at the Adelaide Zoo was an experienced elephant keeper affectionately called 'Our Albert'. The *News* described him as a dashing, tall, good-looking, suntanned Douglas Fairbanks of a man who was originally employed to be the keeper of meat eaters. Born in 1928 in Hamburg as the son of a chief steward on an American shipping line, Albert von Haacht's only ambition was to work in a zoo as soon as he was old enough. When he was 16 he talked his way into a job as a junior keeper at Munich Zoo where he soon collected a scar across his face from a playful leopard. But it was 1944 and in the middle of the Second World War, and when the zoo was bombed he lost his job.

In 1952, several years after the war ended, he took the opportunity to migrate to South Australia. After his arrival he made repeated applications to the Adelaide Zoo, confident he would eventually secure a job there. In the meantime, he took on farm work at Lucindale and on a sheep station outside Whyalla before at last being offered a job at the zoo as keeper of cats. Then he became Lillian's keeper and quickly learned how to handle her, so many hopes were placed on his abilities to care for Samorn.

Unfortunately, he missed her first few days after arrival because he was recovering from injuries received in a serious motor accident. However, given that two keepers were needed to share the workload of looking after one elephant, the other appointed elephant keeper, Derek Bennett, was there to welcome her. As a former London fireman, he took on temporarily the full workload of looking after Samorn and was given a week to learn all about elephant handling from Dave Cody before he returned to Sydney. Derek proved to be an apt pupil, and in less than a week when Albert returned to work, the keepers were able to share the care of Samorn. It was arranged they would look after her on alternate days and weekends. They began teaching her the tricks expected of elephants in this period.

Her arrival prompted a celebration when Samorn

was led to the lawn area to meet a crowd of waiting and very excited children. The children had the opportunity to touch and clamber all over her. She was not even half grown at this stage and, like the small children around her, she just wanted to play and have fun. From the day of her arrival, Samorn was the zoo's star attraction for children as well as adults. Even then, it was often thought, 'What is a zoo without an elephant, the biggest land animal in the world?'

Although she was to be the zoo's star attraction, another drawcard to the zoo for around 26 years was George the orang-utan who arrived there as a gift from the government of Sarawak in February 1950 with a companion called Martha when both were very young. George kept visitors entertained with his many pranks and tricks until his sad death in August 1976. There is now a statue of him within the zoo.

Two well-known celebrities of Radio 5KA, Rex Heading and Lionel Williams, were on the spot when Samorn arrived. They interviewed the Taronga Park keeper and even managed a 'personal interview' with Samorn. When Lionel Williams returned a few days later for another publicity session, he was dressed in a safari outfit with a pith helmet. He also carried a blunderbuss when he was photographed sitting behind Samorn's ears.

Samorn was soon performing several circus-like tricks in public. For one trick Keeper Cody had succeeded in coaching her to stand on a strong two-foot-square (60-cm) wooden box by placing her four feet, each 10 inches (25 cm) in diameter on it. Then Derek Bennett taught her to stand on her hindlegs on the box.

When she so easily mastered a new trick, it was agreed that this 'indicated her high degree of intelligence'.

In these early days Samorn loved to perform and was kept busy with an instant fan club of hordes of children enchanted by their toy-like baby elephant. More importantly, she was a baby elephant they could grow up with.

She had a sense of humour and was naturally nosey so that antics like sticking her head through the cafeteria window to see what was going on were a regular occurrence. She was the quintessential gentle giant baby … always constant … always there … agreeable … and a child at heart just like them. And so the children of South Australia flocked to see her.

Anyone who was anyone wanted to have a say about her or be seen with Samorn, and because it was almost the Christmas holidays when she first arrived, the zoo wanted all the publicity it could get to attract and encourage parents to bring their children to meet her. Samorn, Adelaide's saggy baggy little elephant, prompted a rush in attendances. Journalist Helen Caterer wrote: 'When I visited her … she let me pat her trunk before she squirted water from the tap at me.'

Samorn performed circus tricks every day. And why not? With her as the zoo's undisputed drawcard, everyone was happy. Nearly two years later she gave her first cart rides on 4 September 1958, and though she was intended to give rides with a howdah when she had grown larger, this never happened. Doing circus tricks and later giving cart rides around the zoo for a couple of hours each day, enabled Samorn to earn her keep and benefit from the exercise.

The daily routine went something like this. Between 10.45 am and noon and 3 pm and 3.45, keepers took turns to ride Samorn to the big enclosure for exercise and to put her through her performance. This is what the public expected then. She knelt gingerly, first on one knee, then very gradually lowered her bulk on the other three like a rheumaticky old lady while her keeper slid off her head. She bent her head attentively while her keeper whispered instructions in her rather ragged ear.

Then with the utmost care she put all four feet on a large upturned box. Next her eyes turned to some colourfully painted seats that had been placed upside down in her path. After a word from Bennett she walked up to one and turned it over with her trunk and then, placing each huge foot as gently as a ballet dancer, balanced between the seat back and the rails without touching either. When the performance finished to much applause by the children, they crowded round Samorn to offer cakes and fruit which she accepted with a sensitive trunk, transferring them swiftly to her capacious mouth, especially bananas, her favourite tidbit.

Samorn's official breakfast was plenty of oats, maize and molasses. But she had an insatiable hunger and shortly after her arrival when too many offerings of fruit and sweets caused her stomach upsets, her diet was closely monitored.

Like many children present she was only six years old when she arrived. Many of those children from her juvenile days at the zoo would now be in their 60s. Although Samorn was good natured and put up with a lot of touching and feeling, she knew from an early age when she'd had enough of being fussed over or when the many children crowded too closely and shouted too loudly … for very gently but firmly she nudged them away with her trunk.

Being an elephant keeper can be immensely stressful and keepers found different ways of calming their nerves. In early 1957 Derek Bennett resigned after being found under the influence of alcohol while in charge of an elephant. He was observed speaking to Samorn in a very strange manner and was harshly reprimanding her when she failed to obey his command. Then when Samorn took a mud bath, he waded through the mud to mount her but was thrown off. His odd behaviour was observed over several days before the zoo director, Vince Haggard, asked him to explain himself. But rather than do so, Bennett chose to resign, telling him he could not work with the elephant again.

Perhaps Albert von Haacht was stressed too, when a year later, he was suspected of the same problem. However, as Haggard had brought David Cody from Taronga Park to train both elephant keepers at great cost to the zoo, rather than sack von Haacht he suspended him for a few days. By the middle of the year, on one of Samorn's difficult days, Haggard lamented, 'We badly need a first class elephant man.' He had every reason to air his concerns for as Christmas 1958 approached, von Haacht left. He was replaced by Wally Kubiak.

At the time, the zoo was desperate to hire anyone who knew anything about elephants. But all was not lost, for in January 1959 Bennett was re-engaged on three months probation, claiming to have overcome his problem.

Samorn had several keepers between 1957 and 1963, before her favourite, Hero Nuus, was employed to look after her, which he did for more than 25 years. From 1989, when Hero retired through ill health, a series of other keepers were responsible for her care. Some of the keepers over Samorn's lifetime included Albert von Haacht, Derek Bennett, Wally Kubiak, Doug Cook, Lyn Eglinton, Bill Admundson, Werner Zur Eich, Hans Huemer, E. Huelstrung, Colin Colquhoun, Andrew Mann, Peter Whitehead and, lastly, Paul O'Donoghue. If Hero was the love of her life, Andrew Mann was surely a close second. And there was also another person, John Crutchett, who as a former Lameroo farmer was Monarto's first property manager, and formed a very close relationship with Samorn.

Working with elephants is not only stressful but a physically exhausting form of manual labour. Improving technology has made manicuring elephants' feet less of a chore, but before a labour-saving electric saw was available, Samorn's feet were manicured weekly using a big coarse rasp and an ordinary saw. To see how out of condition an elephant's feet can be, go and see Miss Siam on display in the South Australian Museum. Apart from the care of her feet, Samorn was regularly hosed down and oiled to keep her skin in tip-top condition.

Care for an elephant was not unlike looking after an active toddler; if you don't want it to have something valuable or dangerous, you move it out of reach. The keepers were diligently on their guard to make sure Samorn did not get into mischief or injure herself, but tradesmen working at the zoo were not so careful

'Samorn used trunk as a paint spray.'
Sunday Mail, *22 August 1959. (News Ltd – Newspix)*

and found out the hard way that nothing in the way of tools and equipment was safe from Samorn. Not long after Wally Kubiak became Samorn's keeper, a painter called Ted left his brushes and paint within her reach. Samorn sprayed herself in red paint, even though she was restricted by the chain on one of her legs.

Naturally enough, the brightly spray-painted elephant made the local newspapers with a picture and an article titled, 'You can't see pink elephants at a Zoo but you could have seen a largely red one'. Samorn thought the pot of paint was a treat when the painter left it in her enclosure. Of course being a natural stickybeak, she wanted to know what it was as she stretched over to grab it and drew up a trunkful. Not to her taste, she squirted it all over herself, prompting people nearby to make a dive for cover as they were also sprayed in bright red paint. Samorn, having no idea that her grey skin had turned scarlet, stood by unperturbed not knowing what all the fuss was about.

Keeper Wally was afraid she might blind herself and frantically worked to rub coconut oil around an eye and her long eyelashes covered in paint. She was cleaned down with rags smothered in linseed oil by two keepers who gradually returned Samorn almost back to her natural colour. When it came to who was to blame, the painter denied responsibility, refusing to believe Samorn could have stretched as far as she did to grab hold of the paint pot. He swore she must have broken free from her leg chain or it had come undone.

Like elephants before her, Samorn, 'Adelaide's biggest working girl', was earning her keep. She made over a thousand pounds a year in the early 1960s by giving cart rides to children, an achievement the *Advertiser* poet and commentator Max Fatchen became aware of. For one of his regular articles, *Facing Up With Fatchen*, he had read 'that Samorn, the zoo elephant, raised £1189 in revenue in the past financial year'. He composed this poem:

An elephant will earn her keep
But not a Rocky Mountain sheep,
While tigers lounge with languid claws
And dine with apathetic jaws.
The crocodile with sleepy eyes
In semi-liquidation lies,
The wombat beats a quick retreat,
If someone dares to whisper 'Meat',

The monkey swings from tree to tree
But does not earn a startling fee,
The cockatoos will talk you blind
Financially they're well behind,
Giraffes have never seen a cheque
And why should they stick out their neck?
So who will keep the ledger firm?
The profit-making pachyderm.
With customers like little gods
Upon her honest way she plods,
And when she trumpets loud and clear
Accountants gather round and cheer.
Her income tax was duly sent
Deducting keep and also rent,
The wear and tear on trunk and hide
(See Item. No. 5 … inside).
And when her case has thus been aired
The tax assessor then declared,
'An elephant … that's something new,
'Address her rebate care the Zoo.'

When about 14 years of age, and a teenager, Keeper Wally Kubiak noticed Samorn's teeth needed attention and so the zoo's honorary vet was asked to visit. This vet, Dr P.S. Watts, wrote fondly in the early 1970s of Samorn's teeth problems and her penchant for whisky, prescribed to her for medicinal purposes when it was appropriate. In an article titled 'Twenty years the Zoo Vet' he pointed out that her two regular complaints were her feet 'which pit and need constant attention, the other is her stomach which every now and then rumbles and upsets her'.

On occasions when she had an upset stomach, 'a mild laxative can be got down her in milk, and a bottle of Scotch whisky is then offered and gracefully accepted. I must emphasise that Samorn's taste is impeccable – it must be Scotch – no other alcohol is acceptable and is liable to be rapidly rejected with a rather disastrous result'.

Dr Watts wrote that

> Samorn was perhaps the object of what some members consider my most dramatic effort or perhaps I should say dangerous. It was discovered [in 1963] that one of her teeth had an area of decay. I decided to cut this out and stop it. On the day appointed the president of the Australian Dental Society arrived to prepare amalgam and the operation started. Samorn of course was completely aware of all that was going on and stood still with her trunk in the air. The dental surgeon was discreetly outside the bars! To see properly I had to put my head under her trunk and into her open mouth. After clearing out the cavity about the size of an ordinary tooth I had great pleasure in shouting 'more amalgam' which the dentist kept making while muttering, 'Good God, more!'
>
> When I stood back it appeared that the only people quite unruffled were Samorn and myself. Everyone else was recovering from the jitters wondering what would happen if Samorn ran amok from pain. Since I trusted Samorn completely and she trusted me, we were both quite unconcerned. However, when I left the pen she put on a great turn of fury and I realised that I had left without rewarding her for good behaviour. I quickly obtained some apples and entered the pen and carefully placed them in the untouched side of her mouth. Therewith all was peace and quiet.

In early 1962 the zoo council, for reasons known only to them, decided they would like a companion for Samorn. Fred Basse, the zoo's president, made inquiries through his contacts as to what was available on the market. He contacted Mayfield Kennels & Zoo in Singapore, which quoted £500 for the supply of a young female Indian elephant. But before sealing the deal with them, the zoo director made inquiries also to Major Aubry Weinman, the director of the Colombo Zoo, to see if they had elephants available as it had an excellent name for training them.

Like so many people from the 1950s onwards, Fred Basse found that flora and fauna were becoming harder to obtain. The Major explained 'the export of elephants from Ceylon was no longer permitted'. But at this time there was still a close trading and exchange relationship between zoos and circuses so it was not unusual for the Adelaide Zoo to be approached occasionally as to whether they would like another elephant or if they had one to sell.

Evidently the word was out that the zoo was looking for an elephant. One was offered in February 1962 when the zoo received an urgent telegram from Sole Brothers Circus asking, 'Would you be interested in taking elephant Betty – reply to Bordertown regards Dick Bills Sole – Elephant has killed a man the day before as reported on radio and in press.'

Killer she might be, but by all accounts it was a tragic accident, for Betty had no intention of killing anyone when the circus was pitched at Kalangadoo, to the south-east of Adelaide. The owner of the circus,

John Perry, believed 'that Betty had not intended to kill her old friend, Mr Walker, but possibly in an excess of playfulness, rather than spite, had not realised her own strength'. Sole Brothers Circus recalls that in 1962:

> Bill was working as a journalist for the *Newcastle Sun* and had gone to South Australia to stay with the circus for 2 weeks holiday – he had worked for the circus on a number of occasions ... He went to see the elephant early one morning and it's thought it became excited at seeing him again. It picked Bill up in its trunk and must have knocked his cheek hard (giving him a 'welcoming hug') for a piece of fragmented bone pierced his brain. The elephant laid him gently on the ground and that was how Jean found him. There wasn't a mark on him apart from a slight bruise on his cheek.

The reasons for a captive elephant suddenly killing humans are many, but the circus was giving their 38-year-old elephant Betty a second chance by trying to find a new home for her away from the circus environment. Owners don't necessarily want to destroy an elephant, even when it has killed somebody. Luckily for Betty, a couple of days after the accident she was given a reprieve and was back performing in the circus ring. She had been the main attraction with Sole Brothers Circus for 28 years, having been obtained from the Perth Zoo. More recently, in Auckland in May 2012 a keeper was killed by a zoo elephant, while in August 2012 a young elephant born at Taronga Park attacked his handler, leaving her seriously injured.

Some elephants were not quite so lucky as Betty for there are several well-known cases in the United States where, having killed humans, they were sentenced

The hanging of Mary in Erwin, Tennessee, in 1916. (Johnston University Western Tennessee)

to capital punishment in much the same manner as a human is after being found guilty of murder. Although elephants that killed humans were usually shot dead, an elephant called Topsy was electrocuted in Coney Island in 1903. In September 1916 several forms of execution were considered for Mary, a 22-year-old elephant of more than four tons from the Sparks' World Famous Show. The owner of the circus, Charles Sparks, resorted to what he thought was the best method – by hanging.

Sparks, who had five elephants in his show, instructed his trainers to use 'gentling care' for them.

Mary had been in the circus for 20 years and Sparks and his wife were very much attached to her. But in the town of Kingsport in Tennessee, Mary trampled to death an 'under keeper' who had been hired in a casual capacity when the circus was short of trainers. It was learned after the event that when the temporary keeper had brutally treated her with a bull hook, she retaliated.

Like a human murderer, Mary was arrested by the town's sheriff, and was tied up outside the county jail while her fate was considered. The devastated Charles Sparks, who couldn't afford to take a chance that she might harm a circus patron, decided that she should be hanged. But in order to carry this out, Mary was transported to the railway yards of Erwin, Tennessee, 46 kilometres (29 miles) away, where she was hanged from a massive railway derrick and buried nearby.

As for Betty, awaiting her fate at Kalangadoo, Vince Haggard couldn't make the decision on his own to accept the elephant and he rang the zoo president Mr Basse to discuss the matter. Not knowing the details behind the tragic event but interpreting the telegram as saying the elephant had deliberately killed a man, Basse (as Haggard wrote) 'was very definite that we do not want the animal'. Haggard reluctantly replied to Dick Bills of Sole Brothers Circus at Bordertown: 'Regret we are unable to accept elephant.'

And so the zoo had to wait another year for the elephant it had ordered through the Singapore animal traders Mayfield Kennels & Zoo. Adelaide's new elephant travelled on the ship *Burnside* with four others destined for another zoo; two other young ones were going to the Albertos Circus (Perrys). On arrival, the Albertos Circus elephants were, with the one acquired for the Adelaide Zoo, quarantined for four weeks in the zoo's horse yard, where they were given mandatory tests for medical fitness.

The seven elephants had been supplied by a private zoo in Singapore, run by Herbert de Souza – the same zoo where Lillian in 1934, and then Samorn, had made a stop-over, the latter on her way to Adelaide from Thailand. It appears these kennels not only supplied animals to zoos, circuses and private collectors but, disturbingly, also to medical research establishments. Those were the days when few people questioned the ethics of how wildlife was obtained or to whom they were sold.

Even the Adelaide Zoo in the 1930s was accommodating monkeys in cages provided by the government for the Institute of Medical and Veterinary Science (IMVS), laboratories which were then undertaking experimental medical work into aspects of infantile paralysis (polio) following outbreaks of it throughout Australia.

In papers from a mid 1970s symposium in the USA discussing the use of animals for medical research, one speaker stated: 'In this country trusted recipients still received fliers from places like the Mayfield Kennels and Zoo in Singapore and the Hong Kong Bird Shop.' These fliers, he stressed, often provided 'extra' information with instructions on how illicit things could be ordered and how they would be 'sent through'. The most infamous were the ones in which the Hong Kong Bird

Shop sent full instructions on how to deal with orang-utans they shipped in a double box, with pythons in the front section.

According to the *Straits Times* in October 1962, Mayfield's private zoo, kennel and trading house was Singapore's first when established in about 1918. It was described in 1953 as surrounded by high barbed wire and it lay behind Herbert de Souza's residence on the East Coast Road (now well inland after the 1970s reclamation projects). De Souza claimed he could supply any animal from white mice to elephants. This was long before the establishment of the Babujan Zoo on Punggol Road near Serangoon, that incorrectly fancied itself as Singapore's first zoo.

Arriving at the Adelaide Zoo on 11 February 1963, the unnamed young elephant was the largest of three being transported to Adelaide by the Mayfield Kennels. Already 5 feet 3 inches (160 cms) at her centre back, her girth of 9 feet 9 inches (297 cms) made her almost as fat as she was tall.

After four weeks quarantine in the Adelaide Zoo, Albertos Circus collected their two elephants. Samorn and the zoo's new female baby elephant had met on her arrival and were able to touch each other with their trunks. When the two other elephants left the zoo, Samorn was now joined by the new one.

This touchy-feely meeting was a media event filmed by three TV stations. At this time it was difficult to tell from Samorn's behaviour what she was feeling, whether she was friendly or hostile. Keeping an eye on their first official meeting were the keepers, Wally Kubiak and

Samorn towers over Tara when the two elephants meet for the first time. Advertiser, *19 March 1963. (News Ltd – Newspix)*

Doug Cook. Concerns were that on such occasions a solitary tame beast might object to a new arrival and protect its proprietary right to an enclosure.

Their concerns were unfounded as Samorn showed no signs of territorial behaviour. She appeared happy as she ran her trunk over the small elephant who fitted snugly under her trunk. Samorn indulged in peculiar antics of dominance more characteristic of a male elephant. The younger animal was then returned to her own yard, with repeats of the routine twice more, and spent the night back in her own yard without problems.

No one was quite prepared for what happened the next day when the two elephants were re-introduced. In this short space of time Samorn had become so attached and possessive that it was impossible to separate them. Rather than cause a riot of distressed trumpeting of elephants being parted, the keepers wisely decided that they be left together for the day and night in Samorn's enclosure.

There now seemed no risk in keeping them together. For a week or two it was difficult to separate the animals when it was time for Samorn to be led off to be exercised at the 'elephant ring'. Separated from Samorn, the younger one roared and rumbled in the distance. Samorn's anxiety was such that on returning, as soon as she caught a glimpse of 'Baby' she accelerated her pace so that Keeper Kubiak could hardly restrain her.

Adelaide sought advice from Sydney and Sir Ed Hallstrom and one of the Taronga Park elephant keepers advised hobbles, which were then used to bring Samorn and the unnamed baby elephant under control. The elephants also behaved better when the keepers removed 'Baby' from Samorn's yard and placed her in the quarantine yard before taking Samorn to the elephant ring. When she returned Samorn behaved normally if 'Baby' could not be seen. A few tests were made over several days before she was tried with the cart to ensure that she would not bolt or try to enter the narrow lane between her yard and the rhino's.

Although it seemed much progress was made in training the elephants to suit the keepers' requirements, only weeks after the new elephant's arrival, Samorn's behaviour became unpredictable. On one weekend Samorn 'played up' when the younger one began trumpeting from the yard to which she had been removed, following the usual practice of harnessing Samorn to give cart rides. After she was harnessed, Samorn broke the wooden rails and stepped over the model train railway line into the yard to be with the young elephant. For the time being the keepers could not take a chance with her behaviour and the cart rides were temporarily suspended. This resulted in losses of about £6 a day. By May 1963, Samorn was almost impossible to manage, preferring to be only with her 'Baby'.

At this time when 'Baby' was yet to be officially named, the television channel ADS7 ran a competition to find one. A representative visited the zoo with 648 entries for 'Naming the elephant' to present to Mr Basse the zoo's president. He selected the name of Tara, suggested by three children. A 10-year-old girl of Morphettville, Elizabeth, had written the neatest letter, accompanied by two good drawings with an explanatory note that claimed Tara meant 'beautiful star' in the language of Mysore, India. She was chosen as the winner. Tara was small and fat but Elizabeth still only stood at the height of Tara's eyes.

Shortly after Tara's naming ceremony, Director Vince Haggard, who'd started at the zoo in 1935 as the former Director Ron Minchin's clerical assistant, retired. His replacement, William Gasking, started on 25 June 1963 when Samorn had been there seven years and Tara had been Samorn's adored companion for five months. No one could have foreseen what troubled times the zoo was about to go through.

8

Into the wasps' nest

SHORTLY AFTER STARTING WORK as the new director, William Gasking found he had little time to think about two wayward elephants causing concern to staff, and sometimes the visitors. He was immediately embroiled in a crisis playing itself out within the zoo administration and council. He also found out after he accepted what he thought was a permanent position, that he was actually on six months probation. Not a good start for someone who had sold his house in New South Wales to take up a so-called permanent position in another state.

Gasking took on the role because he wanted the challenge to modernise the antiquated Adelaide Zoo and drag it into the 20th century, even though back in October 1947 the *Advertiser* had written: 'The zoo is now world famous and described by many visitors to be the best breeding zoo in the world.' But only four months into his contract, through the power exerted by the zoo's president, Fred Basse, Gasking was dismissed – one month before he was due to be made permanent. Bitterly, he said, 'I was thrown out on my ear for probationary work that Mr Basse did. I was simply the marionette, Mr Basse's office boy.' He claimed that Basse was doing everything that he, Gasking, should have been doing and he had no idea he was coming to Adelaide to be zoo director in name only. Basse had been on the council since 1935, vice president from August 1957, and president between 1959 and 1964. Alarmingly, his role model for overseeing the zoo was Sir Ed of Taronga Zoo.

One could ask what on earth Gasking was thinking of when he applied for the position as director of the zoo? Yes, he had visited the zoo when he was a boy and his first impression then was that it reminded him of the old Sydney zoo in Moore Park before it shifted in 1916 to its new site on the north shore to become Taronga Park Zoo. There, it was recreated using the latest in zoo design from the Tierpark Hagenbeck in Hamburg, where Carl Hagenbeck was the pioneer of creating naturalistic zoos without bars and fencing, preferring moats.

So remembering how the Adelaide Zoo was, with cages out of the Edwardian era resembling old-fashioned circus ones on wheels, Gasking was having second thoughts. What had induced him to leave the Sir Colin Mackenzie Sanctuary at Healesville, Victoria, where animals roamed in more natural surroundings? He had been there seven years, in which time he saw the sanctuary double in size and almost treble its revenue. One wonders why he left there for Adelaide Zoo. His controversial vision for modernisation meant he flew

William Gasking with his wife outside the zoo residence. Adelaide Truth, *18 January 1964.*

straight into a wasp's nest and unearthed a tradition of shady deals and entrenched cosy arrangements.

As soon as he settled in, Gasking confronted the two major problems that he had to fix if the zoo was to survive in a modern world increasingly anti-circus and anti-zoo. But the worst challenge was the council of the Royal Zoological Society of South Australia; they had no intention of moving out of the Edwardian age.

The first problem he identified was the overcrowding of birds, reptiles and beasts held in a space of only about 18 acres in old-fashioned cages with bars and fences and concrete. Notable was the zoo suffering the stamp-collecting syndrome: the habit of acquiring one of everything, even though some species were not even properly identified.

Indeed, when a particular gazelle died in 1931, Ron Minchin, then assistant director to his father Alfred, recorded that although it had been at the zoo for 18 years, it had never been properly labelled and he refuted it was a Persian gazelle. Young Minchin had intended to send the carcass to the South Australian Museum for proper identification, but was mortified to discover that before he could organise this, his father had already given instructions for it to be fed to the carnivorous animals. He criticised his father for his 'total lack of scientific interest'. And of the type of gazelle it was, he lamented, 'we never will know'.

The other major problem at the zoo was the lack of a functioning drainage system. The solid animal waste was trucked to the Botanic Gardens where it was gratefully received (even though this was found to be contrary to the federal quarantine regulations). The effluent waste-water, however, that was meant to be drained within the zoo, overflowed directly into the River Torrens where it was a health hazard to those who unknowingly swam in it. Gasking was at a loss to understand why the city's health authorities allowed such poor conditions to exist. And in the zoo itself, where the terrain was mostly flat, heavy rain in winter turned many enclosures into unhealthy pugholes making it an unsightly health hazard for the animals.

With modernisation at the top of his agenda, Gasking had the choice to either dispose of half his exhibits and focus on quality, or look for a larger site. The latter was not a new idea; the issue of a new site for the zoo had been raised in the local press in November 1952 and later in December 1958 when a site at Greenhill was recommended by the town planner, Mr Stewart Hart. However when Gasking raised the issue in 1963, he hit a raw nerve.

All too quickly Gasking realised he was trapped in a difficult predicament. The zoo was in a desperate need of modernisation, yet what was the point unless he could install better drainage, have a new plan of the zoo drawn up and get rid of half his stock? This would double the space for the remaining species and he'd get rid of bars and wire to create a naturalistic environment with the use of moats.

The writing was on the wall for him as soon as his ideas were made public in the *Advertiser.* He couldn't avoid an interview with the press for, of course, the journalists were eager to hear what the new director's plans were for the future of the old-fashioned zoo. And naturally, Gasking couldn't help voicing his view of acquiring an alternative site somewhere in the country. Ironically, when it was first established, the zoo had been granted 62 acres (25 hectares) of Section 443 in the Hundred of Noarlunga at Mylor, but it was never used. The land was later transferred to the Stirling Council by the state government. Gasking aired his zoo plans as a long-range program, while stressing the present zoo site could be used as a subsidiary attraction if the zoo moved to a new and bigger site elsewhere.

The next day the angry Mr Basse vouched that the Zoo Society's council did not share Gasking's personal opinions. Further, the council supported the status quo, including the large number of animals.

Gasking spent his first day in the office, and the

mornings of the rest of his first week with the outgoing director Vince Haggard meeting the staff and finding his way around the zoo. Things started to go wrong when a number of snakes suspiciously went missing on the third day. Then a few weeks later, in early August, two Commonwealth police officers paid a visit to question him about the consignment the zoo had sent to the Colombo Zoo in mid July.

They informed him that while the zoo's accounts were in order they were probing into the affairs of the zoo's agent, Gerald Hargrave. The son of a former Adelaide lord mayor, Hargrave worked for John Skipper of Avitraders Ltd based at Gays Arcade, Adelaide, since about 1955. When Skipper went to live in Sydney in 1960, Hargrave became the Adelaide manager of Avitraders Ltd, which then traded in flora and fauna.

The Adelaide Zoo had been dealing with Avitraders since at least 1953, buying and selling wildlife, exotic and native, although this was not then an exclusive arrangement. After the restriction of the export of Australian fauna that came into force in December 1959 (when their business was severely cut and they were unable to pay outstanding debts to the zoo), the company sought a contractual arrangement with the zoo. They met with the Society's president and director in August 1961 to discuss becoming the zoo's *bona fide* agent for the export of fauna overseas on a commission basis of ten per cent of stock value. The appointment of Hargrave was agreed verbally, according to Fred Basse's later evidence in court.

The Commonwealth Customs Department brought

16 charges against Hargrave and Avitraders and the case was heard in court over 13 days from 2 September 1963. The crux of the case was that the documents Hargrave submitted to customs to cover the consignment did not match the zoo's. In an article titled 'Bye bye birdie' in the *Bulletin* of 4 July 1964, it was stated 'the Adelaide Zoo had allowed its name to be used for what in effect was a purely commercial venture, and, one would imagine, a very lucrative one'.

It was certainly lucrative for Sir Edward Hallstrom, director of Taronga Park, for it was revealed in court that he had used his Adelaide Zoo life membership to ship two consignments of soft-bill birds through Avitraders to an agent in the USA. Hargrave explained that after Hallstrom's request to export the birds they were shipped on behalf of the Adelaide Zoo.

The police claimed Hargrave had also taken the opportunity, whilst agent to the zoo, to add many extra pairs of parrots and cockatoos to zoo consignments. He was convicted of making false customs declarations in respect to bird exports under Section 234 of the Customs Act, and his verbal appointment as the zoo agent was expressed, during the summing up, to be 'as loose as the wind'.

Hargrave's activities with the zoo had been spotted by a diligent quarantine officer who had noticed an increase in the number of fauna being exported within a very short period. When the Chief Quarantine Officer stressed the prohibition of exports by dealers and the limiting of exports to *bona fide* zoological gardens only, Basse and Haggard – and later Gasking – were required to explain in minute detail to government authorities the zoo's more recent export transactions. The zoo's administration books were seized by the Crown Solicitor for further scrutiny and to be used in the court case.

Hargrave and Avitraders were found guilty on eight charges but given overly light fines. Magistrate Wilson explained that 'although what Hargrave had done amounted to a breach of the Act ... to stray outside the limits of his authority as an agent of the Royal Zoological Society', Hargrave and Avitraders had not made any deliberate defiance of the Customs Act. However, they still stood to suffer and were given the minimum penalties. In his conclusion, Magistrate Wilson stated that the tremendous costs of the hearing were 'out of all proportion to the enormity of the offence'. However, many in the know would have disagreed with his justification for downplaying the seriousness of the offence. It is possible the magistrate had no idea just how extensive the illegal trade of wildlife was in South Australia, and this case was just the tip of the iceberg.

Gasking also had to deal with the delicate situation of a certain Belgian national, a Mr A.P. Saabe, who claimed to be the director of the Parc Marie-Henriette in Ostende, as well as representing the Societe Zoologique de Belgique. Saabe was initiating clever deals to procure large consignments of native birds to Belgium despite the prohibitive regulations of the *Flora and Fauna Act of 1959*.

Months before Gasking took up his duties, Saabe's activities to circumvent the federal regulations came under scrutiny as he attempted to trade with the Perth,

Adelaide and Sydney zoos. He'd already received a consignment of Australian birds in early 1962 after a meeting with Fred Basse, the Adelaide Zoo president, and Vince Haggard, director, John Skipper, director of Avitraders, and his agent for the zoo, Gerald Hargrave. So pleased was Saabe with this consignment that Fred Basse said every effort would be made to supply his needs. However, in mid 1963, when Saabe was under surveillance by Australian Customs, Basse distanced himself from his 'illegal activities'.

The Commonwealth Police and the Australian Quarantine and Customs departments' intention to prosecute Saabe as a wildlife smuggler had the potential to also expose Basse and Hallstrom's own suspect activities. They needed to do something drastic to save their necks. In early June 1963, only days before Gasking took up his duties at the zoo, quick action by Commonwealth authorities prevented Saabe leaving Perth by Qantas with over £8000 (equivalent to $120,000 today) worth of Australian birds – but only because Basse and Hallstrom dobbed him in for his illegal activities. It was believed Saabe was 'acting on behalf of black market interests overseas'.

Unfortunately, it was while these sensitive matters of possible wildlife smuggling activities by the zoo's agent were under scrutiny that Gasking took up his duties. From the first day he and Fred Basse clashed. Gasking not only met with antagonism and obstruction, but because the council was loyal to Basse, the members were reluctant to make decisions or improvements not vetted by Basse and they kept their distance.

While Gasking was employed as director to manage the zoo, he was never allowed to actually get on and do so. Instead, Basse undertook activities without telling Gasking, such as having an old male puma shot or inviting VIPs to visit the zoo without Gasking's knowledge. Basse, without telling Gasking, gave instructions to labourers to hurriedly build temporary stalls for a consignment of new beasts from Taronga Park because the Adelaide Zoo had not been forewarned of their arrival until the day before.

The consignment may have been one of Sir Ed Hallstrom's many acts of generosity, a sweetheart deal perhaps – or were strings attached? Basse and Hallstrom were like two peas out of the same pod. They ran their respective zoos as though they were private property and only they knew what was best for them. Basse's obituary in the 1970s stressed his very close friendship with Sir Ed.

The *Flora and Fauna Act of 1959* made it almost impossible to export Australian wildlife. When poor legislation made the ban a form of 'prohibition', the wildlife trade simply went underground in a harmful way, creating a huge and profitable black market that has continued unabated ever since, and which now uses 'sophisticated techniques and international criminals'. Informed aviculturalists and ornithologists were concerned that a valuable opportunity was lost to legislate for a regulated trade that would have seen the cost of the most desired species drop to sensible prices and have seen a fall in mortality rates that would otherwise lead to extinction.

The Exhibition Committee of the Adelaide Zoo had also recognised that the Act of 1959 would be too restrictive and had requested the Commonwealth quarantine authorities as early as 1958 to modify restriction, but without success. It is noted that after the passing of the Act, 'soon international bird trafficking to and from Australia was centred on Taronga and Adelaide Zoos'.

In 1963 it looked as though Basse was involved in such underground activities himself when Gasking stumbled across a possible shady deal concerning his attempts to procure a cock lyrebird from Victoria, illegally. When Basse realised he was implicated, he tried to extricate himself from it the best way he could, but it didn't fool Gasking. Some of the Society's council members weren't too impressed either when they came to know about the lyrebird saga, and they started asking pertinent questions. But it would seem that another reason Gasking was dismissed was that he would not cooperate with the lucrative 'bird trade'.

Always looking for 'true' but sensational stories, the *SA Truth* newspaper quoted Gasking: 'Since I've been appointed I've had phone calls asking people to exchange birds. When I told them this was a contravention of the quarantine laws, they said they have always been able to do it.'

Cosy habits were entrenched, for as early as 1943 Basse was housing zoo birds in his own aviary at home because 'they would have better chances of breeding'. This practice was often repeated. In August 1961 the zoo rented cages for about 24 birds at the home of Ron White, the zoo's carpenter.

Quoting Dr Peter Reeves, the journalist Stewart Cockburn wrote 'with the knowledge of certain members of the zoo council, trappers and traders had been allowed to use the zoo as a "boarding house" for birds. Trappers sent their birds to the zoo, where they were looked after until it was convenient to export them in the zoo's name, thus evading customs regulations'.

The lyrebird episode concerned a batch of several letters written by a 'J. McNamara' of Berwick, Victoria, who offered a cock lyrebird to the zoo in a covert manner. The letters were so concerning that when Gasking discovered them he took the originals from the file to the council member who was the government representative, Dr Peter Crowcroft of the South Australian Museum, for discussion. The original letters were removed to make copies and missed from the zoo's files by Basse, who now had every reason to worry the secret was out.

Gasking couldn't really know if Basse was involved in an illegal transaction or not, for Basse craftily wormed his way out from the matter with a perfectly plausible explanation. He stated that 'many practical jokes are played on the zoo each year, sometimes by letter, but in greater numbers by phone, which our office staff will confirm'. Basse wrote 'in itself this letter could only be regarded with suspicion and after explaining the position to two other Council Members, I decided I was going to follow it up on the lines suggested by the writer to see what would eventuate'.

There was no way of knowing about the extent of Basse's involvement in the matter, or whether it would have become a form of entrapment had the deal gone ahead. Not convinced of Basse's ploy, however, the Society's council member Percy Pollnitz criticised the president 'for pursuing the matter while assuming it to be a hoax'.

And what was happening with the elephants at this time? Only days after Gasking became director, the elephant keeper, Wally Kubiak, lost his cool with a visitor who had berated him for beating Samorn with an ankus (pick-like device) near her exercise yard in the sight of the visitor's children. The woman complained that he was rude and used the words 'bloody Australian' and 'mind your own business'.

Kubiak told Gasking that the complaining woman spoke rudely to him, calling him a 'bloody New Australian'. He told her the elephant had to be controlled and advised her where to go to report the incident if she didn't like it. He denied swearing at her and stressed that he was always careful of his speech and actions when near members of the public.

Suspension from his duties while inquiries were made had a grave follow-up for Kubiak. Soon after, Gasking received a phone call from the police that Kubiak had been arrested for driving under the influence and was sentenced to 100 days in prison and five years' disqualification from driving. When his son turned up at the zoo at a later date to collect wages owing to his father, it was also to inform Gasking that his father had resigned.

Samorn and Tara's poor behaviour continued and in October 1963, despite Basse and Gasking being at war with each other, Gasking brought to the notice of the Society's council members the danger Samorn caused by frequently swinging her trunk at children as they played on the barrier rails encircling the elephant ring. He proposed the area be enclosed with a chain mesh to minimise the danger.

Gasking was constantly criticised by Basse for not noticing what was going on around him, but Gasking stressed his lack of confidence in the keepers' abilities to maintain control over the elephants. It must have been a surprise when Basse endorsed this view, blaming the keepers for paying too much attention to the adulation of the visitors. The keepers were advised to give their fullest attention to controlling the elephants and to avoid long chats with visitors, who tended to divert their attention from the animals. He suggested the two keepers desperately needed further tuition from one more experienced than them. In the meantime, the keepers' prime task was to maintain a stricter control over their charges and aim for an improvement overall.

It was about this time that Hero Nuus was employed, even though he knew absolutely nothing about zoo keeping and even less about elephant handling.

One Society council member was concerned about visitor safety generally and suggested Tara be sent off to a reputable circus to be trained properly, especially as the two keepers confessed to losing control over her. The keepers thought the elephants should have separate quarters as Samorn was now dominating Tara and there was not enough room for two growing elephants.

As the elephant problem worsened along with the Basse conflict, Gasking requested a crisis meeting with several members of the council with Basse present to resolve some of the friction between them and what was going on in the zoo generally. It was to no avail.

With cart rides temporarily abandoned, Samorn was exercised with an empty cart for some weeks while Tara was not exercised at all. By this time it had transpired that Kubiak had been bullying Doug Cook, destroying his confidence to take the elephants out for exercise. But now that Kubiak had left, Lyn Eglinton, who claimed some experience with elephants, reassured Gasking that he and Doug Cook could restore control over Samorn and Tara between them, and with the help of new keeper Hero Nuus.

It was during such confused times that, on a bright spring day, a lack of attention seriously affected the wellbeing of not only the elephants but the rhino as well. Perhaps no one had noticed how hot and sunny it had become when the hides of the beasts were oiled. They suffered from sunburn and their skins peeled, and Basse blamed Gasking for the oversight.

The tension levels went up another notch after a small child was severely frightened when she turned her head to find Tara standing beside her. Perhaps until that very moment the little girl had never seen a real live elephant and was totally unprepared for its awesome size. She screamed in sheer terror and Tara, spooked by the screaming, took flight and bolted back to her enclosure leaving a trail of debris behind her in a way that only a half-ton terrified animal could. It was a zoo incident that reached the newspapers.

The *Advertiser* reported that Tara startled visitors

when she broke loose while being exercised near her compound, and bolted. To the squeals of excited children, keepers pursued the trumpeting elephant as she tore around flowerbeds and raced past the zoo kiosk where people were enjoying afternoon tea. The bellowing Tara was caught as she headed back towards her compound, Samorn adding to the din with loud encouragement. A visitor from Clare said that children ran in all directions as Tara scurried past.

The children in the vicinity thought the incident was part of an elephant act and ran behind her screaming excitedly. Tara arrived back at the elephant enclosure to Samorn's bellowing and trumpeting. When Doug Cook caught up with Tara to calm her down, Samorn would not move back from the gateway to let Tara in, and the noise of the children continued to add to the overall pandemonium.

Harsh action though it was, the two keepers struck Samorn with canes to force her to move back to allow room for Tara to enter the yard. But several visitors, believing what they saw to be an outright case of animal cruelty, reported this incident to the media. One woman complained to Gasking of the 'brutal treatment' of the elephants. No visitor knew the full story of why this had occurred, but judged that Samorn was being beaten for no obvious reason.

Both elephants were spooked for days. Samorn swiped dangerously at children with her trunk alongside her walkway if they swung on the rail or ran across in front of her, and during the night she smashed up her water trough. Tara, tense, unpredictable and difficult when people were about her, was exercised outside of zoo hours. When neither elephant could be exercised their behaviour deteriorated. Samorn's temperament was no longer docile, and Tara trumpeted a lot when Samorn was giving rides or being exercised. The tension increased daily as neither could bear to be separated from the other and bellowed, trumpeted and rumbled as though they would never see each other again.

In the early 1970s, Trevor Klein was looking at an exhibit opposite Samorn's yard when Samorn was being walked to the exercise ring. He didn't move when he felt an elephant's trunk looping around him. She was looking for something in his pocket. He remembered 'for the first time feeling threatened and uncomfortable at the zoo' as she stood right next to him.

Older than the little girl terrified by Tara, Trevor was nevertheless concerned. But by his own admission, he loved going to the zoo as a small boy because among other things some of the wild animals did provoke in him an element of fear; for others it was a combination of 'excitement, fear, awe, sadness and nostalgia, with unease about the captivity of animals'.

Tara had created a minor incident by running amok within the zoo but an elephant on the loose outside a zoo can spook a whole town or region.

In August 1954, during the night, Dolly, the elephant of Wirth's Circus, escaped from her circus train truck at Cook on its way to Woomera in the far north of South Australia. She was not missed when the train departed without her. Hours later, when the train reached Watson 60 miles on and circus authorities were informed that their elephant was causing havoc in vegetable gardens back in Cook, did they find the truck doors wide open.

Residents at Cook were soon made aware of Dolly's presence as her loud trumpeting and the barking of dogs awakened many people who were too scared to venture outside. When the local stationmaster went in search of the source of the commotion he walked into Dolly at the side of his home where she had tucked into the few precious bushes from one garden and then cleaned up a patch of a dozen cabbages in another. Both startled, she took off along the railway lines.

Meanwhile, circus attendants boarded the next train going westwards. It stopped when they encountered Dolly on the railway line. But as soon as she recognised the circus attendants she took off at great speed and 'went bush' for more than three miles before she stopped and they caught up with her. She was then walked back to Cook where she made another dash for liberty when she saw a freight train.

The attendants were able to persuade a driver and his motor truck to join in the chase and Dolly was eventually caught and put on the freight train. Within minutes of the train leaving the railway yard, she had 'trunked' open the pins in the locks and pushed open the doors. This time she was prevented from making her escape when two of her legs were chained to the truck floor and all doors were wired. After a delay of more than 40 hours, the freight train was finally underway on its journey to Pimba, where Dolly was returned to the circus.

Nothing quite as exciting as that happened to young Tara. She didn't escape from the zoo, but ran amok within it causing minor damage, her actions all part of these troubled times.

Less than two months into Gasking's directorship, Basse and his loyal supporters of the zoo council were clearly worried they could not rely on Gasking's support to continue zoo activities in the way Basse wanted.

A clever little plan was hatched. Sir Ed Hallstrom arrived from Sydney to give specific advice at a Society's council meeting held towards the end of August 1963. Hallstrom's advice to operate Adelaide Zoo along the lines of Taronga Park – that is, managed by one man only – made Gasking redundant. The motion to be dealt with at the next council meeting was as follows: 'In order to honour the President (Mr Basse) for his distinguished services and to qualify him for election to the International Union of Zoos Directors, and on the advice of Sir Edward Hallstrom, it was resolved that Mr Basse be President and Governing Director.' The motion continued by way of confirming legitimacy that 'it was further resolved that, notwithstanding the previous resolution, the Director is responsible directly to the Council for management of the Zoological Gardens'. The motion was deferred until the next meeting with every intention of it being legitimised. However, as events at the zoo moved swiftly into an ever-deepening crisis, there was no need to vote for the resolution as the crisis resulted in the intended outcome anyway – which was the getting rid of Gasking.

It was amidst this unpleasantness and chaos that the famous African author, Joy Adamson of *Born Free* fame, visited the zoo in October, confirming in the minds

Joy Adamson with Elsa at Meru, Kenya, circa 1959. (Elsa Conservation Trust)

of many locals their concerns about the zoo being antiquated. She was so distressed about the confined quarters of some animals, particularly the larger mammals, that her criticisms were blasted across the local newspapers. Gasking recalled that she was unable to understand why the zoo did not use an area it still owned in the hills at Mylor. He recalled that 'the sight of those big bears in little cages was one of the things that upset her ... she said that anybody who kept animals had an absolute duty to keep them in the best conditions'. Gasking agreed that 'those are my sentiments entirely'. The thoughts of this international celebrity mattered, to the consternation of the zoo council, for the publicity triggered off a barrage of similar complaints to the newspapers by disillusioned past visitors.

9

Ringing the changes

THE ZOO SAGA REACHED ITS CLIMAX at the end of October when Gasking was handed a signed letter by Basse and two council members, M.T. Phillips and Dr A.H. Lendon, informing him that his probationary period would not be made permanent and that he was relieved from all duties on receipt of the letter. No reason was given for the zoo council's decision because Basse would not supply one. But Gasking was certain that the reasons included his statements about transferring the zoo to another, more spacious, site. Other reasons were that they wanted to gag him from making further comments about the zoo's cramped conditions or any suggestions he might make of its possible involvement, or particularly Basse's involvement, in illegal trafficking of native wildlife. He wrote in his journal: 'I am disappointed that my efforts to improve conditions in the Zoo have not been appreciated, due I believe to my having uncovered evidence of illegal & dishonest activities on the part of some of the senior council members. If this decision stands, I feel very sorry for my successor.'

Not happy at the way Gasking was treated, three brave musketeers in the Public Service, the dissenting zoo council members of Dr Peter Crowcroft from the SA Museum, Percy Pollnitz, known as the 'big boy' from the SA Tourist Bureau, and Noel Lothian, the amiable director of the Botanic Gardens, approached P.H. Quirke, the Minister of Lands, to set up an inquiry into the affair. Writing about 'The Zoo Affair' in January 1982, the journalist Stewart Cockburn recalled that Crowcroft had stated their action had 'been forced on us by the ruthless and brutal behaviour of a small coterie of elderly egotists on the council'.

Of course the government wanted to know, too, what was going on, seeing they had intended to grant £42,000 to the zoo over the financial year. To find out they appointed the Auditor General, G.H.P. Jeffery, and the Director of Lands, J.R. Dunsford, to head an inquiry into the organisation, management and operation of the zoo before any further funding was transferred to it.

On 13 November 1963, the *News* quoted Sam Lawn, the Labor MP for Adelaide: 'From what I have read it appears that Mr Gasking has some ideas for improvements at the zoo ... Just because a man has ideas for improvement is no reason to have him shot out of the organisation. Surely we don't want to continue doing things as they have been done for the past 100 years.'

The next day the *Advertiser*'s most well-known cartoonist Pat Oliphant (who went on to be a much-

Advertiser, *14 November 1963.*
(Courtesy Pat Oliphant and Newspix)

admired Pulitzer prize winner for his cartoons in the United States) nicely captured the dissent at the zoo. Portraying the three wise monkeys, the cartoon had a caption that read 'I guess it's no good asking them what's going on around here'.

Gasking complained that since his appointment Basse had antagonised him and was the 'leading light' in his dismissal, backed by two other council members, Lendon and Phillips. He said, 'They tried every inducement to get me to resign, but I said I didn't want to.' He continued, 'There was no doubt that Mr Basse virtually runs the zoo turning up everyday of the week for most of the day. He started his day in the office for a short while and then went all around the zoo. The staff didn't like his daily attendance and most of them didn't know who was the boss.' Gasking also said: 'When he [Basse] found faults with the zoo instead of consulting me, he brought up questions and wanted explanations when the committee met … he was very much possessed, not unlike … in a dictatorship, behaving as though it was all his, carrying out all the directorship work from foreman to animal feeder.'

This negative criticism and serious behaviour problems with Samorn and Tara were wearing out the staff, and matters were not helped when Doug Cook left for a few weeks after a serious motorbike accident. This left the new but inexperienced keeper, Hero Nuus, who'd only been at the zoo since September, to do the best he could with help from other keepers.

In a confidential report, written by Basse to cover up his own suspect behaviour, he condemned Gasking as having a lack of general zoo knowledge and of non-Australian mammals in particular. However, Gasking had been successful at the Healesville Sanctuary, and after leaving the Adelaide Zoo the state government appointed Gasking to investigate the establishment of a park like Healesville. This became Cleland Park in the Adelaide Hills, which he managed with much success.

What stuck in the craw of the 'loyal' zoo council members was that in February 1964 they were required by the government to obtain an area suitable for a sanctuary for the purpose of holding animals, surplus to the immediate requirements of the zoo, with the council then voting 8 to 5 for the 'former director of Healesville Sanctuary' to manage it. Sir Keith Angas said that 'employment of this man went against the grain in the light of his past behaviour, but it might be in the Society's interest eating humble pie'. The risk was that the state government might withhold its annual funding to the zoo. While Angas was pragmatic, the other dissenters saw this action as comparable to blackmail. Among those opposing Gasking's appointment were his most formidable foes, namely Basse and Lendon.

Two months after Gasking's dismissal, a special meeting was held in January 1964 at which the action was reconsidered. Council member A.G. Owen-Smyth doubted that any of this unpleasantness would have occurred had Gasking been appointed [permanently] while Peter Crowcroft confirmed what many had suspected, that 'fundamentally, troubles existed prior to his arrival'.

Council member Sir John B. Cleland moved an amendment 'that Mr Gasking be given a further trial period of six months but that the President take six months leave with Dr Watts acting as President'. But while three voted for this amendment, eleven others voted against it so that Gasking lost out anyway, despite the best efforts to have him reinstated. 'Bluntly', stressed Dr Crowcroft, 'these difficulties would remain unsolved whilst the President remained in the chair.' Adding that the Society was neither a successful organisation nor a profitable business, he rubbed salt in the wounds by declaring that it also had a 'bad reputation'.

After Basse was requested to step down as president in January 1964 (but remained on the council), it was stated that the trade in birds dropped to a tenth of what it had been two years before. Only one Society council member (other than the government-nominated ones) had ever bothered to meet Gasking during his period at the zoo. While the overcrowded ark had been rocked violently in a rough sea during the fiasco, it miraculously survived a sinking.

When the government inquiry was completed and the report made available to the Society's council members they were far from happy and decided they 'would receive the report but not accept it'. Three members claimed the government report was impertinent and even libelous in light of criticism contained within it. Percy Pollnitz from the South Australian Tourist Bureau felt the future of the Society's council was dependent on a much-needed and essential change of heart and attitude as indicated in the report, but as things were, he believed, there would be no change.

Sir Keith Angas reminded members that accepting the report was to infer acceptance of criticism. And naturally enough, while the Society's three government council members accepted the report, 11 of the members wouldn't and three others abstained. Loyalty to Basse remained unshaken until the day he stepped down as president. Distancing himself from what had gone before, when Dr P.S. Watts took over Basse's position in early 1964 he let it be known that 'the policy of the zoo will be that of the council and not that of the president'.

When William Gasking was dismissed, the previous director, Vince Haggard, was invited to return to the zoo briefly as superintendent until a new director was appointed. The irony was that the zoo had stepped back into a brief state of suspended animation under the same director who took over the reins in February 1940, during the Second World War, when it was impossible to update or build anything.

Richard E. Minchin with his family, circa 1890s.
(Royal Zoological Society of South Australia Inc.)

For 60 years since the Society's establishment in July 1878 (and opening to the public in April 1883) the management of the zoo had been controlled by just three directors. Curiously, they were three generations of the Minchin family. The name was so synonymous with the Adelaide Zoo that Leo Oosterweghel, current Director of Dublin Zoo and a former bird curator at the Adelaide Zoo, remembered a story about a small child who on visiting the zoo was most disappointed at not seeing an animal called a 'Minchin'.

The first of the family involved was Richard Ernest Minchin. A useful senior public servant with no training in zoo management he was made director of the zoological gardens in November 1882. He'd been on the original committee and at 41 years was probably

Alfred Coker Minchin. (State Library of South Australia B 28220)

the youngest when he helped form the South Australian Acclimatization Society, later to become the South Australian Zoological and Acclimatization Society. He made substantial donations towards the new zoo from his extensive aviary in North Adelaide. Several years later a gracious residence was built for the zoo's director, and here the three generations of Minchins built up their reputations. When Richard was absent on collecting trips in Southeast Asia, his second son, Alfred Corker, who also made his career in the zoo and was made assistant director in 1891, acted as director. When Richard Minchin died in 1893, 36-year-old Alfred with his hands-on experience was the obvious candidate for the position, despite the many worthy applicants. He effortlessly slipped into becoming 'a most competent authority' who it was claimed 'made a paradise at the gardens'. Moving further up the social scale of respectability than his father, Alfred was voted

Ronald Richard Minchin, 1938.
(State Library of South Australia B 27878)

in as a member of the Adelaide Club and held court at the zoo on Sunday mornings.

Alfred was director for 41 years. Both Alfred's sons were also keen about zoology and wanted to run their own zoos, but it was the second son, Ronald Richard, who joined the zoo staff as a 19 year old in 1923. As with the grandfather before, when Alfred died in 1934, Ronald, who had been assistant director since 1929 when he was only 25, stepped into his father's boots to become director. He brought to his position a passion for Australian birds, parrots in particular, with the help of his curator of birds, John Manfield, who he claimed 'was the best keeper we have ever had'.

Alfred Keith Minchin managed the Koala Farm, circa 1940s. (From the collection of Nancy Maddern)

Between them, from 1930, they bred rare Australian parrots with financial support from Dr William Anstey Giles, which enabled them to erect the first parrot breeding cages. From November 1935, support in this specialised area came from two new Society council members, Fred Basse and Dr Alan Harding Lendon, the latter a surgeon and aviculturalist. Their combined efforts led to the zoo's great reputation for its bird collection. When Ron Minchin became director he wrote in 1935 that the Adelaide Zoo Gardens had 'the unique distinction of being the only institution in the world to exhibit the whole family of Grass Parrots'. Sadly, Ronald was in the job as director for only five years. He died of cancer at the age of 36 in February 1940.

Ronald's older brother was the controversial Keith Minchin, who in 1927 set up his own animal enterprise, the first commercial venture allowed on the Adelaide Park Lands. It was known as the Snake Park, and he expanded it in 1936 when it was renamed the Koala Farm after he included koalas and other animals. Jovial, dominating, argumentative and publicity loving, Keith was a vice president of the Adelaide Zoo from 1935 until the menagerie's closure. While rarely attending meetings (he was a Society council member until about 1945), he fully supported the zoo and helped his brother with labour and materials whenever they were requested, for the zoo never had enough of either. His hands-on help was despite outright disapproval of him by several of the council members. When Keith Minchin closed his Koala Farm in 1960, most of the animals, reptiles and birds were donated or sold to the zoo along with carriages and carts.

After his brother's death in 1940 it appears some of Keith's goodwill towards the zoo evaporated for he billed the council for all the work he had done for the zoo over the years, for which he was mostly reimbursed.

When the management by the Minchin dynasty came to its end in 1940, the zoo was locked into old ideas and attitudes, hindered by the effects of war and the ravages of the Great Depression when funding was in short supply. Although the Depression was well over when Vince Haggard became director in 1940, the zoo hadn't recovered and there was little funding for updating and modernising.

Relief came with small improvements after the Second World War when Sir Edward Hallstrom of Taronga Park and a member of the Adelaide Zoo for some years donated much-needed funds to erect bird-cages. As stated in the official zoo history by C.E. Rix, *Royal Zoological Society of South Australia 1878–1978*: 'Such beneficence must have seemed like manna from heaven to the council and members of the Society and in recognition of his generosity, Mr Hallstrom was made a vice president.'

So, in 1964, while Vince Haggard was acting director, the zoo stepped briefly back into a no-man's land until a new director was appointed. The zoo, long trapped by economic restraint, was further stifled by the lack of vision, particularly in four of the Society's council members who, when they died in the early 1960s, had collectively served for about 150 years.

William Edward Lancaster was appointed director in April 1964, and with changes to the council membership major new plans for the zoo's future were proposed, despite them being mostly Gasking's ideas anyway. The 1965 development plan included basic services of deep drainage and electricity, the conversion of Edwardian buildings to functional designs, the moating of previously fenced enclosures, a children's zoo, reptile house, walk-through aviary, Australian section and new enclosure for apes. The zoo was dragged into the 20th century by vigorous implementation of the plan, but it hadn't been the only Australian zoo condemned for its poor conditions. Taronga Park Zoo was denounced as 'filthy and revolting' and people were 'severely critical of the one man control being exercised by Sir Ed Hallstrom' until his retirement in 1966.

During this critical time the Adelaide Zoo battled

on with their two dysfunctional elephants, but 18 months into the job William Lancaster believed that keeping two elephants presented problems and that Tara was more a liability than an asset. To make matters worse, the problem was compounded when Keeper Doug Cook wanted to be taken off the job. This was no surprise for months before, in pursuit of an escaped deer running about in North Adelaide, he fell from a motorbike, fractured his skull and was away for two months recovering.

Cook's decision provided further problems, for he would be difficult to replace, and so Hero Nuus of only two years zoo experience was dropped in at the deep end to take control of Samorn and Tara. Zoo life could be less stressful with only one elephant and the former director Ed McAlister hints at the rumours that Hero fostered restlessness between the elephants to expedite the zoo seeking another home for Tara.

Other zookeepers were used as 'fill ins' to help Hero Nuus with both elephants, but by October 1965 the situation was so taxing that the majority of the Society's council members were in favour of selling Tara. Further, if two elephants were to be retained the present accommodation would soon become too small for the growing animals. Apart from these crucial facts the on-going cost of feeding two elephants was a major consideration.

Before and during Gasking's four months at the zoo, Samorn and Tara's bad behaviour was a major issue. But despite it being so, Gasking was too overwhelmed by the other serious problems that had come to light to worry much about them. The elephants were affected by the tense atmosphere, as were the staff. The keepers were confused about who they should take their orders from, and it is more than likely the elephants were affected by their anxiety.

The period between 1963 and 1965, which saw the arrival of Tara and shortly after of William Gasking, was an extremely difficult one. Gasking had been keen to modernise the antiquated zoo, the Society's council hostile to change. This inevitably created a toxic mix that brought the zoo's many problems to the scrutiny of the government, media and public.

10

Samorn and Tara

THE ARRIVAL OF TARA in 1962 was wonderful for Samorn and all the visitors who came to see the two juvenile elephants at the zoo. But it was not always wonderful for anyone working at the zoo, or especially with the elephants. Overnight, the place became several decibels noisier as the two boisterous elephants communicated as only elephants do, by trumpeting, rumbling and bellowing. Samorn, once docile, quiet and passive, changed. Suddenly she had company of one of her kind, an elephant to love. And it was as though Tara encouraged Samorn to come out of her shell and show her true character. Remember Samorn was an adolescent and demonstrated the kind of challenging behaviour known in human teenagers.

This was in the days when elephants gave rides and were expected to perform and put on a good show. This was a fairly easy thing to do when there was just one elephant being dominated by keepers, who could also choose to have a close relationship with her. After all,

Tara with children from Gilles Plains Infant School. Advertiser, *6 May 1963. (News Ltd – Newspix)*

they were all she had. This no longer applied once Tara arrived.

Maybe the zoo staff did not understand the dynamics of communication between animal and keeper and how they changed when another elephant was introduced into the zoo community. As the zoo found out, having two elephants was not simply twice as much work as one.

In a herd environment in the wild, female elephants have very close relationships within a matriarchal hierarchy. The most senior cow is leader. When male elephants reach puberty they leave the herd to strike out on their own or join 'bachelor herds', returning briefly at breeding times to mate when mature. For the newly captured wild female elephant, the dynamics of interaction are changed for her when she is placed within a zoo or circus. While she will be closely connected to other elephants when there is more than one, a group of females is usually dominated by a male keeper. Moreover, in more recent times groups of elephants in a zoo or circus are usually made up of females, as they are easier to control.

As with most animal species in captivity, including elephants, boredom is the enemy of an otherwise healthy and peaceful way of life. Needing constant work, play, training and exercise to avoid this, elephants look to one or two people to control them but also to give them care and affection. The dynamics of relationship are even more crucial when an elephant is the only one in the circus or zoo, for its keeper has to fulfil many roles. Not recognising the role of dependency or the trust that goes with this responsibility might see a decline in the temperament of the solitary elephant that could lead to otherwise preventable accidents or even death.

It appears that until a captured female elephant is almost fully grown, it is docile, gentle, affectionate and controllable. This changes, no matter how careful the keepers are with their management, when the elephant approaches maturity and becomes moody, cantankerous, unpredictable and increasingly reluctant to perform or do tasks. It might also be that their long memories of past negative issues, for years internalised, are suddenly vented in unpredictable and dangerous forms of expression.

Ivor David Balding is an American circus owner. His

documentary, *One lucky elephant,* about his performing elephant Flora, pointed out that 'after 18 happy years, things changed. Flora wouldn't do tricks and started to act up. She wasn't happy ... [she] used to love being around humans, but now she'd go to her trailer and close the door behind her. I just knew she was fed up with the circus. She needed to be an elephant ...'

Samorn was remembered as being predictable during her earliest years at the zoo before Tara came. In hindsight the surprise is that this period lasted as long as it did, for three of her first five keepers were fond of the bottle and did not always treat Samorn kindly. Several times when they were under the influence of alcohol and lost control of her, they resorted to harsh treatment, which led to unwanted media publicity and a loss of job for a couple.

In fact compared to the docile and older Lillian, Trevor Klein remembers the keepers treating the feisty Samorn when she was young with aggression – and not necessarily mild aggression. 'I saw her being hit with the handle of a tool. I have seen an ankus jabbed into her with its pointed end just above the hook that goes behind her ear and I've seen that hook rammed down over the back of her ear.' While Samorn did not seem to hold a grudge at the time, her earlier experiences of keepers smelling of alcohol, or their harsh treatment, saw her have occasional outbursts of lashing out at visitors, children, keepers and other zoo employees when she was older. Several who knew her said that in later life she was hostile to anyone who smelt of alcohol or those she simply didn't like.

It is likely the zoo management hadn't thought through how the dynamics would change when Tara arrived in 1962. Samorn, no longer on her own but now with her own 'baby' to care for, was no longer solely dependent on her keepers, Wally Kubiak and Doug Cook. For the few years Tara was at the Adelaide Zoo, she reminded Samorn that she was indeed an elephant and she wanted to behave like one with her new companion. The elephants were the major draw-card at the zoo in this period because they gave cart rides and did circus tricks. Giving rides was a rich little earner of several thousand pounds per year, it literally paid for their keep. The zoo could not afford to lose this kind of revenue nor could it take the risk that the paying public might have been just as satisfied to see two young elephants playing and acting out together, even if there was a lot of bellowing and trumpeting.

The director, Haggard, followed by Lancaster, and the elephant keepers used what knowledge and experience they had about elephant behaviour in captivity to try and understand the interaction that occurred after the arrival of Tara. This was when the discipline of ethology, the study of the behaviour of animals, was still in its infancy, as was exotic zoo veterinary science. Had they known and understood more and adjusted the way they exhibited their elephants, there would have been no need to sell Tara to the circus.

As it was, because elephants are so needy, there were several elephant keepers for the two elephants. The show went on as before with both elephants performing for the public, as the zoo believed that was expected of them.

In the zoo's history, ponies and camels have also given rides. There were also circus performances that were hugely popular when they were introduced in May 1939 with monkeys riding dogs and orang-utans riding bikes. This followed the employment of F.M. MacFarlane, an animal trainer, who was paid more than other keepers to use his skills for arranging a miniature circus similar to those put on in other zoos. The zoo authorities firmly believed there was nothing that could be considered cruel in the acts, despite several complaints by the public that led to an investigation by the Royal Society for the Prevention of Cruelty to Animals in 1942. The RSPCA believed that animal acts such as these helped to perpetuate the public demand for circuses, thus entailing unnecessary cruelty in training animals for the purpose. But despite the criticism the entertainment continued. After MacFarlane left seven months later, the zoo continued the act of the orang-utan performing on a bicycle on alternate weekends.

Twenty-two years later, when William Lancaster was appointed director, the situation remained the same. And so, after a further two years of the anxiety of managing two temperamental elephants, Lancaster decided Tara was more a liability than an asset and had to go. He and the council came up with reasons they should get rid of her, and used this claim to sell her.

She was offered to the Auckland Zoo, and they were interested following the death of their star and first elephant, Jamuna, at 48 years of age. But when Derek Wood from the New Zealand Zoo visited Adelaide, he decided Tara did not meet their requirements and it was

1968 before his zoo replaced Jamuna with one called Ma Schwe.

All was not lost, however, for Sole Brothers Circus and Wirth's Circus was touring locally and heard Tara was on the market. When Sole Brothers offered £1250 she was sold to them with the consent of the Exhibits Committee. She would have made a wonderful Christmas purchase, as the circus collected her on 24 December 1965. Curiously, for all the fuss the elephants caused when separated, Samorn showed no sign of fretting when Tara left. But then Hero was there to pick up the emotional pieces – if there were any to be picked up.

Sadly, selling Tara led to her untimely death in a traffic accident in 1985, when her circus was travelling between tent stops in Queensland.

Although Samorn showed no signs of distress when Tara left, her character had changed after having a close companion for almost four years. Although there were no outwards signs of grief, perhaps it was internalised, never to be forgotten. Or is it possible that she was secretly pleased to see her go? It may be because she was older and more mature, she now had 'attitude' that became part of Samorn's personality in later life.

She needed her one-to-one relationship with Hero Nuus when Tara was sold off. While another keeper may not have been single-minded in caring for Samorn, Hero was.

11

Samorn's Hero

DESPITE ALL THE NEGATIVE REVELATIONS in 1963 and 1964, under the leadership of the visionary William Lancaster, 'a veterinarian scientist with a very good text book understanding of animals', the zoo faced a bright future. Innovations were introduced that ensured the zoo had a continuing role in the local community over the next few decades. It also redefined itself to become more active and relevant in the worldwide zoo community by concerning itself with the reality of conserving and breeding endangered species in captivity, and setting out to educate the public, which included providing a Children's Zoo.

It was also a period when the days of the zoo being run by members of the Adelaide 'Establishment' were effectively over. When the zoo was first established, the council was made up of the most influential and often wealthiest 'gentlemen', with little or no knowledge of running a menagerie, let alone a zoo. Zoological societies were not necessarily formed to study animals in the academic sense. Rather, the precursor to the Adelaide Zoo was an acclimatisation society in which the original council sentimentally and pragmatically introduced the furred and feathered inhabitants of their European homeland so they could hear familiar songbirds and hunt animals they knew for pleasure, and pursue economic interests.

The original council had looked to their 'junior' working member, a public servant, Richard Ernest Minchin, to organise and manage for them such an enterprise. This he did as well as he could with his limited knowledge of zoo management picked up on the job. With immense dedication and enthusiasm, which attracted a growing local patronage, the little zoo joined the ranks of those in many major cities in the Western world in the 19th century. A zoo was necessary if Adelaide was to be viewed as civilised and progressive.

But in the early 1960s this notion, along with other out-dated practices, was being severely challenged. In September 1963 a Dutch gardener, Hero Nuus, began working at the zoo and soon after became Samorn's keeper. From the little village of Oostwold in the north-east of the Netherlands, he and his wife arrived in Melbourne as migrants in June 1959 aboard the ship *Willem Ruys* under the Netherlands–Australia Migration Agreement (NAMA). When he was first employed by the zoo he knew nothing about elephants but with a 'can-do' attitude he quickly stepped into the breach when the zoo found itself without anyone who could handle not just one elephant, but two.

This began a new period for Samorn, for Hero was

Hero Nuus trains Samorn to perform, circa 1960s. (Photograph courtesy of Frank Manfield)

there for her over the next 25 years, sharing 'a mutual devotion'. He went on to become an 'old school' keeper who was an immensely important person at the zoo solely because of Samorn. He built his life around her in his role as her keeper. Only a handful of keepers could manage an elephant, but for those who did, being in charge of a several ton animal, the largest in the zoo, gave them a sense of power and a degree of importance.

Shortly after Hero began work, Director Gasking left, to be temporarily replaced by the former director, Vince Haggard, until William Lancaster was appointed in April 1964. As a British Colonial Service veterinarian, Lancaster had been involved in animal quarantine activities in Thailand and Malaysia. When he decided to 'retire and be lazy', he chose to migrate to South Australia. According to the former director, Ed

'Lanky' Lancaster and Samorn. News, *30 June 1977.*
(News Ltd – Newspix)

McAlister, he contacted the zoo president Dr Percy Watts and during an informative meeting he asked, 'Where does a vet get a job around here?' Watts realised the type of man he was dealing with and responded by inquiring, 'How do you fancy being the director of our zoo?' And that was it, according to Ed, 'for he became the next zoo director'.

'Lanky', as William Lancaster was called, was often seen dashing around the zoo on a bicycle. He is fondly remembered by another past Dutch employee, Leo Oosterweghel (now director of the Dublin Zoo), for wearing a funny suede jacket with patches on the elbows. He learned that, as well as having an 'upper-class' English accent, Lanky could swear terribly in Dutch as he had been in a prisoner of war camp with

Dutch prisoners (and on the Burma Railway). It was Leo's lucky day when he applied for and was appointed to a birdkeeper's job on one spring day in 1976, because Lanky believed the Dutch made good zookeepers.

Within Lanky's first year a development plan was drafted, with valuable input by the council member Dr Alan Lendon. It supported Gasking's vision of replacing the bars and fences in favour of moated yards. By Christmas of 1965 a Children's Zoo was opened; the restaurant, that had been leased out for years, was taken over by the zoo; and the zoo's first but small souvenir gift shop was established. In this period, incredible as it seems now, the average visitor spent the equivalent of only about 20 cents within the zoo, apart from the admission fee.

In Lanky's third year the zoo sold Tara to a circus. Samorn, now on her own again, had her dedicated keeper Hero all to herself; the man who became widely known as the Adelaide's Zoo's 'elephant man'. He comforted her through what could have been a difficult time after the loss of her companion, and he recorded 'when dealing with an elephant [the trick] is to keep talking to let her know you are in a good mood. I tell her little stories and talk eye to eye. Then she knows I'm feeling okay and when I'm in a good mood, she is'.

In July 1967 the third of the zoo's moated enclosures was completed, for Samorn. It was much larger than the others and included a pool and a heavy duty rubbing post.

In 1969, six years after Hero began working at the zoo, Werner Zur Eich, one of the 'new breed' of keepers, was appointed Superintendent of Mammals. Werner remembered that when he first took up the job all the keepers looked twice his age and yet he was expected to be their boss and tell them what to do. He was glad he had already decided to grow a beard (which he has never shaved off), to make himself look older and more authoritative. He needed his best diplomatic skills to avoid rubbing anyone up the wrong way as he gradually introduced new methods in animal husbandry.

One of his first impressions of the zoo was that there were too many animals in small spaces. Amid this overcrowded area of cages and cemented enclosures there were only about four or five major trees, with few others to create much needed shade for the animals, or along the pathways. This made for an open, hot and dusty place for visitors throughout the long summer months. Although the zoo council had considered planting more natives and exotics, nothing serious happened until Werner's arrival. He set about changing this over more than three decades with the valuable surplus plant donations he received via Rex Kuchel, the zoo council member from the Botanic Gardens.

Further, as self-seeded trees were strategically re-located throughout the zoo, it was slowly transformed into an open but shady woodland, despite there not being a plant budget in this period. This was 1968, when the entrance fee that had remained the same for eight years was raised (allowing for a change to decimal currency in 1966), to 40 cents (from 30 cents) for adults and 20 cents for children.

From this time and for more than 20 years, Hero

and Samorn continued as an affectionate double act that eventually attracted the attention of a zoo member wanting to write something of their close relationship and Samorn's antics.

Many stories about Samorn and Hero were retold in 1986 in a children's book by Victoria Raine. She wrote *Elephant in our Zoo* through Samorn's own words. The stories captured her high-spirited personality and her wit through the many incidents that her handler, Hero Nuus, conveyed. It seems Samorn acted up whenever the keeper's back was turned. Her curiosity was such that, unable to leave well alone, she created many incidents, very much like a naughty child.

Hero Nuus and Samorn. News, *10 August 1966.*
(News Ltd – Newspix)

Werner Zur Eich, Superintendent of Mammals, with Samorn.
Advertiser, *1 November 1992.*
(Photographer Grant Nowell, Newspix)

Even in her own house, if any fixture could be loosened, broken or dismantled, Samorn would have a go. When she repeatedly played with the old original double timber doors of her house until she finally jammed them, a new set in quarter-inch steel was made that could be locked firmly open or shut so that Samorn could no longer fiddle with them.

Anything left near her enclosure was fair game too. Werner knew Samorn's obsession for gadgets, especially those that could bounce or make interesting sounds when played with, until smashed to smithereens. Werner lived in the supervisor's house at the zoo for more than 20 years and had an emergency buzzer right next to his bed. One night he was awoken by the din of something being hammered and bashed. Knowing it could only be Samorn, he nipped down to her house in his pyjamas to find her having a smashing time with a wheelbarrow in her trunk. She had pulled it into her enclosure and bounced it up and down until the top part had been detached from the base. With its air-filled tyre giving it extra bounce, she deconstructed it with some gusto.

Anyone who worked at the zoo has their own hearsay version of the following event involving Dr Percy Watts, who (like his father before him) was, amongst other positions, an honorary vet, in his case from the mid 1960s onwards. Watts was in the habit of parking his prestigious dusty-blue Mercedes-Benz car within the zoo precinct close to the rear of the Elephant House and next to the Animal Hospital. This, as the story verifies, was an unfortunate choice for it was parked next to a mound of vegetables and fruit collected every few days from the Central and East End markets for Samorn and other beasts.

Sometimes, when Samorn's enclosure was being cleaned, she was allowed out as a special treat, to forage for choice tidbits from the mound. One day while Hero's supervision was diverted for a few precious seconds, Samorn, who had already spotted the Mercedes parked nearby, saw an opportunity to try out her weightlifting prowess and test whether the car would bounce. In the blink of an eye it was already too late. Hero turned to see Samorn with her trunk through the open window, enabling her to lift the car as she tilted it onto just two wheels to test the suspension.

Caught in the act, Hero bellowed, 'Drop it, Sam,' and being the good obedient elephant that she was, she dropped the car. Although it bounced, the *Bendigo Advertiser* for 8 December 1986 reported that 'it bounced straight up under her trunk and hit her on the end of her sawn off tusks that dented the roof of the car'. But the tale didn't end there. When Dr Watts drove his damaged car to the garage for repairs, so the story goes, he was already a little intoxicated (for he was a man who liked a drink). Naturally enough, when he was asked how the damage was caused, and being somewhat tiddly, he said it was done by an elephant. One could not blame the garage mechanics for thinking he was having them on.

Much is made of the zoo elephant's capacity to make money by giving cart rides to earn her keep, but Samorn also earned the zoo extra revenue by showing off her talents in advertising. Her first stint was in March 1961, when she was part of an advert for Waldorf carpets.

With this work experience under her harness, many years later in August 1987 Samorn was given the starring role in another carpet advert. This time the gimmick was for Samorn to test Dupont's stain resistant carpeting. Unexpectedly, it became big news, capturing the imagination of the local and even national media. Samorn proved more than equal to the task when she was given the opportunity to 'test' the new carpet in her own house.

Samorn tests the Waldorf carpet. Advertiser, *11 March 1961. (News Ltd – Newspix)*

The event was covered by SAS Channel 10 *Eyewitness News* three nights running, and by the Adelaide *News* and Melbourne *Sun* newspapers along with 'personal interviews' over every radio station in Adelaide. For some time Samorn continued to appear in Le Cornu's adverts for Dupont carpeting wherever it was sold. For the zoo's trouble in organising the promotion, they received a donation of $800.

Birthdays of some of the more personable zoo animals, apart from Samorn, were celebrated. For the hippo Newsboy's 30th birthday in January 1962, a special cake of vegetables was prepared by the staff and photographers of the *News*. Samorn had several celebrations, and because they were organised as major events, they always attracted hundreds of children and adults along with the media.

When Samorn was 21 years old in 1971, Werner organised a big birthday party for her. He had calculated her age from her actual day of arrival at Adelaide Zoo on 30 November 1956 as a six-year-old. He wanted to use this opportunity to do something special but it was also a reminder to the public to come and visit the zoo. So he planned her party as a media event to be held just weeks before the school summer holidays began.

Being such a big animal, Samorn needed a big cake. Werner designed and constructed two large fake tiers out of plywood. The first tier was four feet square and three feet high. The second tier was three feet square and two feet high and was painted white with pink trim. And finally for the real cake, the top layer was created as an English 'bread-pudding', made from bread, sugar-buns and some fruit, like bananas, for extra flavour. In eight elephant-bite-size pieces, this layer was stacked solid to a total of two feet square by one foot high. When sugar-icing didn't work on the very moist recipe, Werner and his wife sprinkled shredded coconut and added 21 pink 'rock-sugar' candles to complete this gigantic wonder.

On the chosen day of Samorn's birthday, the glorious-looking creation was set up on the lawns in the middle of the 'elephant ring'. To add to the atmosphere for the birthday girl and the crowds of children present for the occasion, Samorn was joined by her many animal friends with their keepers. These included a koala, echidna, hand-raised monkey, python, kangaroo joeys and various other pets from the Children's Zoo. And 'of course', reminisced Werner, 'around those hordes of excited children and on the outer ring, were the invited media with cameras snapping'.

As for guests at a wedding waiting for the bride, the moment everyone had waited for came as the star of the day slowly plodded up the path. Samorn entered the arena to loud cheers and she was presented with a very large silver 21st birthday key. This didn't interest her one bit, she had already spotted the enormous cake.

Believed by the many children to be forgetting her manners, she sidled up to the six-foot-high cake and stretched out her trunk to touch and sniff. Then with one swipe all the sugar-candles were collected, and gobbled in one go. Following this, with a couple of karate chops using her trunk, she had the cake in messy pieces, to the squealing delight of the kids. All that planning, design and creation vanished in a blink of an elephant's eye and the children cheered and shouted. It had been one enormous success where the star of the show shared not one slice of her huge cake with anyone, but ate the lot by herself. And everyone was happy.

Samorn had her funny if not taxing moments when language was misinterpreted, for English, German and Dutch were spoken around her. On one occasion she completely misunderstood Keeper Peter Whitehead: when he was noisily clearing his throat she took it to be a command. Hearing a scuffle behind him, he turned to see the several tons of Samorn standing on her head in the sandpit. He realised it was likely she had heard the throat clearing as a command in Dutch as known only by her former keeper Hero Nuus. Peter was perplexed for he had no command to bring her down from this precarious position. Trying to make eye contact with him through her long lashes, when at last she realised there was a problem, she returned onto her back feet and then pushed herself up on all fours and moved back into her normal position. She was quickly returned to her enclosure and given extra herbage for her meal that night.

Until 1982, Samorn's life at the zoo was one of routine, with plenty of activity and exercise that included the cart rides. However, various incidents with her cart had caused concern about the children's safety, along with her being fond of swinging her trunk at anyone within reach while giving rides, which could also cause serious injury. Indeed, on an occasion just before the elephant rides were abandoned, she was playing up with a keeper who was unable to keep her under control. The cart was at a precarious angle and full of frightened children, and the situation about to develop into a serious incident had not Werner Zur Eich happened to come upon the dire situation. From a distance he bellowed at Samorn to 'Stand still!'. At first she did not respond and he yelled at her again. When she saw it was Werner giving the command, she froze on the spot, the cart returned to its horizontal position, and everyone sighed with relief.

In another incident Samorn reached over a guard-rail and grabbed a pusher. It didn't contain a baby but, such

Hero Nuus and Samorn giving cart rides in 1973.
(Photographer Gwen Jones)

was the concern, that another rail was bolted higher on to the fence to prevent it happening again. A sign had been put up some time before warning the public that she grabbed bags.

While she wanted to perform or pull carts that gave her the valuable exercise she needed, she had to be carefully controlled due to an increasing number of incidents. Growing fears that her unpredictability could result in a serious accident led to elephant rides being ceased permanently in August 1982.

While Hero was confident he could control Samorn, the other keepers felt he was being overconfident, and so the director's recommendation to discontinue the

Samorn takes a bath. News, *8 January 1968.*
(News Ltd – Newspix)

rides was, with great reluctance, agreed to. There had been many benefits in giving cart rides, apart from the revenue earned from them, for it gave Samorn valuable exercise, which she needed, and stimulation and close contact with zoo visitors, whom she loved. When the rides ceased she had to be exercised outside of zoo hours, and a program of new forms of stimulation and activity, to overcome her repetitive behaviour of rocking side to side, was put into action.

Sixteen years after Tara was sold, the idea of obtaining another female elephant as a companion for Samorn was considered and her enclosure was further enlarged to almost double the size, with a donation bin placed by her enclosure. And in a spirited gesture, one of the zoo members donated a thousand dollars towards purchase of another elephant. But this never happened.

The latter part of the 1970s was a troublesome time for the zoo. What was the zoo to do when the attendance figures dropped to an all time low? It became a priority to publicise the zoo at this time for attendances had plummeted by almost 40,000 per year, and by then the lucrative practice of selling peanuts from the kiosk to feed the animals was discontinued. One answer was to decorate the perimeter walls facing Frome Road with gaily coloured murals of the zoo's residents and in January 1980 send the director to publicise the zoo and its new sponsorship program on the popular TV daytime show – that had been running since 1968 – 'Touch of Elegance'.

Unable to take Samorn, the director took a tame boa constrictor instead. Unfortunately, he was in the event unable to introduce it to the audience as the mere presence of the reptile in the Channel 10 building caused a mass exodus of all female employees, including Jaye Walton, the hostess of the show. (This was the same Jaye Walton who much later, as Thai Consul, helped this author search for Samorn's origins). Such was the display of terror that the studio audience and the viewers never saw the snake. But the sponsorship program did go ahead to great success and years later, in 1991, the Charlies Bar at the Hilton International Hotel sponsored Samorn on her 40th birthday.

After advice from consultants in 1984, it was decided there should be a single body managing the affairs of the zoo and the Zoological Society should have a new management board comprising just 12 members instead of 19. In a period leading up to this decision, when there was a shortage of staff and they were having to work long hours overtime, they went on strike several times for better pay and conditions with Hero Nuus as their shop steward for the Federated Miscellaneous Workers' Union.

The strikes at the zoo created particular hardship because animals still had to be fed, despite a stoppage of work. The situation was left to the Superintendent of Mammals, Werner Zur Eich, who ensured they were fed. Rallying to the cause in June 1980 and in March 1985, a dedicated group of workers made up of Werner's family and members of the zoo council, along with their wives and relatives, fed the animals.

One solution to find more staff to work directly with animals was to discontinue the jobs of the driver who picked up sick, injured or dead horses and the slaughterman who prepared them for the carnivores. This was achieved in 1985 when packaged meat was bought instead.

Samorn and many of her zoo companions had strange new hazards to deal with in the mid 1980s, such as emergency helicopter landings on the university's oval opposite the zoo and then the carrying of patients by ambulance to the Royal Adelaide Hospital at any time of the day or night. With the helicopter often roaring very low beside the zoo, Samorn and many other animals went berserk, with some of the more timid injuring themselves. In October 1985, with the commencement of the Grand Prix car race in the city, part of the entertainment was the aerobatic display by fighter jets that included bomb bursts as they dived low over the zoo, causing much distress to the animals. Although the director successfully lobbied for no repeat performances, the situation was different later with the emergency helicopters that flew in to the Royal Adelaide Hospital's helipad from all directions at least once a day to save a life.

Unimaginably, one night in early 1985 there was nothing to protect the animals when two youths stealthily broke into the zoo and killed or maimed 65 of the smaller animals that were mostly tame and vulnerable. While the offenders admitted it was planned, there was no known motive other than sadistic vandalism. Among the tame animals in the Children's Zoo, special little relationships had been formed, particularly between old-timers like Cynthia the llama and Josie the western grey kangaroo. Also killed were well-known animals that children were able to touch in the Children's Zoo, such as Anna the Indian antelope and Sukey the kangaroo joey.

The atrocity shocked the public. Following the disbelief and sadness that people were capable of such a barbaric act, donations flooded in towards helping to replace many of the animals, along with a valuable donation from the company BP.

People remembering Samorn just before she was transferred to Monarto, had differing opinions about her welfare. Of more concern to the then director, Ed McAlister, was that when news leaked out that Samorn was to be transferred, he was besieged by angry grannies

Samorn is given a scrub down as part of the Scouts Bob-a-Job Week. News, *15 May 1975. (Photographer Chris Mangan, Newspix)*

who accused him of depriving them of being able to easily take their grandchildren to see the elephant in the city. After this kind of harassment he worried, along with the Curator David Langdon, what it would be like for them if Samorn died 'on their watch'.

12

Sharing the love

WHEN 54-YEAR-OLD HERO RETIRED due to ill health at the end of 1989, Samorn felt neither abandoned nor neglected. Her future without him had already been seriously considered. Other keepers she could relate to and trust were introduced throughout the 1970s and 1980s. The day he retired was possibly sadder for Hero than Samorn. When he left after 25 years, the *Advertiser* declared 'their marriage … has had its ups and downs. Good and bad times …'.

So by the time of Hero's retirement, Andrew Mann was well trained to care for Samorn. It is more than 20 years since he relinquished his job as her keeper. Not looking much older than a man in his mid 30s, he is quiet in manner and gives the impression of one who is extraordinarily patient. In the 1970s he was fresh out of university with a degree in botany and zoology, and was looking for work when he heard the zoo needed an extra hand to replace a staff member on sick leave.

He enquired and was employed the next day as a casual native mammal keeper for two weeks. With his foot in the door, he was then re-employed as a bird keeper, followed by a position with the ungulates: that is, animals with hoofs.

He worked around the zoo where needed, and whenever he walked past Samorn's enclosure he yelled out to her, 'Good day, Sam.' In this way she became comfortable with him in her environment. Their relationship became closer when he assisted Hero preparing Samorn's food and with such chores as cleaning her enclosure.

Werner, the Superintendent of Mammals, had watched Andrew for some time and was convinced he was the right person to handle Samorn. He noticed she always reacted positively when Andrew was about, making the type of rumbles and squeaks elephants make when greeting someone they like. Knowing there were few keepers capable of handling Samorn, he invited Andrew to consider the job but warned it could be a stressful position. When Andrew agreed, Werner and Hero began training him for the task in the way they felt would be best for Samorn after Hero retired.

In the late 1980s, when it became clear that Hero would not reach the conventional retirement date, due to poor health, Samorn's sole dependence on him was gradually reduced by the training of several keepers. Once she was comfortable in their presence, they were able to handle her, feed her and clean her enclosure without fear.

After Hero retired – and sadly, he died comparatively

Andrew Mann with Samorn. Advertiser, *29 November 1990. (Photographer David Mariuz, Newspix)*

young in 1991 – Samorn bonded very closely with Andrew. Her love for him would show particularly when walking with him. Andrew's wife Donna noticed that 'she'd swagger with her tail swinging'.

For Andrew, however, it was not all beer and skittles working with Samorn. Now a fully matured elephant of about 40, and without the companionship of other elephants for a long time, she was strong-willed and liked her own way. She was stubborn, had moods, tantrums and off days – just like a human. Her past interaction had always been with people, so she wasn't like a herd animal and she wasn't like other elephants. More importantly, she was neither just another elephant, nor a pet.

When Andrew accepted the challenge, he went through the drawn-out process of convincing her she had to treat him as leader of the herd. Convincing Samorn she was not the boss and there was a pecking order she had to submit to, took time and patience. But Andrew stressed that with a human adult weighing between 70 and 80 kilograms on average, you could not forget for one moment that Samorn was a few thousand kilograms heavier; a wrong move could prove fatal.

Although she appeared docile and good-natured most of the time, Samorn was known to take an instant dislike to certain people, especially those who smelt of alcohol or showed an aggressive manner toward her. Andrew confirmed that on several occasions Samorn 'had one of her naughty moments, her incredibly naughty moments'.

So it was that a new employee arrived to work with the team. As Andrew stressed, 'Sammy wasn't your average elephant and you needed to familiarise yourself with her before going too far with her'. She did not appreciate a new keeper bossing her about in an authoritarian manner or using harsh physical punishment for disobedience. Further, no sane and sensible person who knew anything about elephants would consider entering Samorn's enclosure or her sleeping area with her there until she had shown signs of accepting them.

It was immediately obvious when the elephant met the new keeper that Samorn simply did not like him. But he had experience with elephants and believed he would be all right if he worked in a certain way around her – with Andrew there. On one occasion the new keeper came into the Giraffe House where Andrew was cleaning and told him: 'I want to go in with Samorn and check her feet and various things.' Andrew advised him that he himself hadn't been working with Sammy for long and wasn't too sure how she would react to another person. But the new keeper dismissed Andrew's advice, assuring him it would be OK, stressing: 'I've worked with elephants, I know what I'm doing.' And so Andrew was reassured, despite his initial unease. What followed next belied belief.

Andrew recalled that when they entered the enclosure they gave her some food before lifting and checking all four feet for stones. They then went out to clean the yard.

'We were talking when Sammy came up to us and stood near us but facing the new keeper. I thought this a bit strange but noticed she had finished her food and was probably seeing if we had more. That was when she hit the new keeper. In a matter of a split second he was on the ground and Sammy was trying to head stand on him.

'Fortunately he fell to the ground where the walls met, otherwise he would have been killed when stood on,' said Andrew.

But Samorn hadn't finished. Just as she was about to have another go and crush the keeper, Andrew yelled at her in his best reprimand voice, commanding her to back off. He said, 'She completely froze in her tracks and looked at me as though to say, "Oh shit," and then went into a state of shame and remorse by making tiny squeaks like a mouse, as she often did.' Perhaps she felt like a naughty child caught out when about to break a valuable toy. Meanwhile the new keeper got up and moved behind Andrew, who had put his own life at risk.

The injured keeper was taken to hospital for X-rays of a suspected broken collarbone, but luckily all he suffered was bruising. When the zoo management considered how best to deal with the problem, there were two options. The new head keeper could continue working with Samorn and throw his weight around using 'old elephant styles' to make a point of domination over her. If he chose this option he risked possible repercussions and leaks to the media.

Sensibly, the new keeper chose to not continue working with Samorn. Andrew concluded: 'Samorn was like that. She was careful with those she respected but was not that careful with those she didn't.'

Soon after the incident, an emergency exit was cut through the back wall of the elephant house. This meant the house could be cleaned without direct contact while Samorn was in the yard and the big door closed into the elephant house. An alteration more than 70 years overdue for an elephant house built in 1900. In *The Noah's Ark Syndrome*, C.F.H. Jenkins wrote 'the most powerful but often the most docile of zoo inmates are the elephants ... [but] because of their enormous strength and bulk even a slightly fractious elephant can be highly dangerous and so usually elephant enclosures are provided with several doors or slits through which the keeper can escape at the first sign of trouble'.

Not long after, Samorn attacked again. A zoo maintenance man who worked in a noisy and disruptive manner upset the animals wherever he was. On a day when Samorn had had enough of him and his attitude, she head-butted him. Although a big strong man who was involved in martial arts, he was lucky to crawl away from her as fast as he could without injury.

During this period, the zoo was looking to develop land at Monarto obtained in 1975. Under the management of just two men it was developed throughout the late 1980s as an open-range safari park of 1000 hectares where large and mainly African animals would live in a herd environment.

The property surrounded by low farm fencing had been used as farm paddocks for agistment and annual crops, but without much success. The crops were continually ruined by mobs of emus and invaded by rabbits and foxes.

Ed McAlister, the incoming director, focused on Monarto in 1991 when Samorn was relocated there and a 15-kilometre feral fence installed under a Commonwealth employment scheme. The site was developed to provide open areas for exhibits and was heavily planted to provide shelter and an attractive environment to encourage native birds and animals and to allow for the regeneration of native flora.

Towards the end of the 1980s, many concerned visitors complained to the Adelaide Zoo and wrote letters to the editor about the way Samorn was enclosed in her small environment at the city zoo, with not a lot to do. Although much time and effort was made to keep her busy and active, after she no longer gave cart rides her displays of rocking side to side in repetitive behaviour attracted the attention of visitors genuinely concerned about her welfare. Even the zoo records reported in January 1991 that she was 'looking depressed, not eating this morning. Better in evening'.

The decision to transfer Samorn to Monarto had been

made months before Ed McAlister took on the directorship in July 1991. Although Samorn was an Asian forest elephant from Thailand, it was decided she would benefit from being transferred to the open plains of Monarto where she would live out her remaining years. It was hoped that it would eventually be in an elephant herd environment that would also give her what she'd probably never had before, lots of space. Considering Samorn's 35 years in a city zoo with a passing parade of visitors, it is debatable whether this was a good move. With just two staff based at Monarto to oversee the 1000-hectare property, Ed McAlister was pleased to announce an increase by 50 per cent in one month when he appointed Graham Goldsmith as 'the third man'.

Earlier, in about May 1991, it was reported 'Samorn is receiving training from Andrew Mann in preparation for her transport to Monarto'. Once she was moved there he would 'stay with Samorn at Monarto for four months'. Every day for many months she was led into the large crate that would transfer her there and left in it for almost an hour, about the expected length of time the journey would take.

She was transferred in November. As she was going into her container she momentarily hesitated, having had second thoughts but, to everyone's relief, she then trundled in without fuss. Extra precautions were taken that day as a nervous several ton animal might panic when the journey was underway. Taking no chances, the staff travelling with her to Monarto were fully armed. This rather shocked the new director, Ed McAlister, who had to face the realities of the situation.

He accompanied her to Monarto and recalled it was a very anxious elephant that emerged from the crate 'a foot high in elephant dung'. He was awed by the experience of standing close to an elephant for the first time. Nothing had prepared him for the experience of placing his hand on her craggy and coarse-haired body to help pacify her after the 70-kilometre journey.

It was stated in the annual report about the transfer to Monarto in November 1991: 'It should not be entirely overlooked that we are fortunate in having an elephant whose personality and behavioural health has allowed the job of relocation to be conducted considerably safer and easier than it might otherwise have been.'

It was far from an easy settling-in period for Samorn after transfer to Monarto. Having been used to her compact city zoo environment, she encountered the extreme of large open paddocks where both animals and zoo visitors were sparsely scattered. Initially, Samorn was filled with dread of the open spaces, an agoraphobia of sorts. From where she was located, she couldn't see another animal, nor was there at that time a passing parade of visitors, and there was only three keepers for company. It took Andrew several months to settle her in and familiarise her to the new and open environment. For the first week or so, he slept overnight with her, so there was someone familiar in her company. The comforting sounds and smells of the city zoo gave way to the strangeness of the new country ones. However, after a period of time, Andrew gave her more space and went home to Lobethal each night.

It took Andrew a month to coax her into leaving her

shelter. Too afraid to explore, she had no idea about a newly created mud wallow at the far side of her two-hectare paddock. Andrew persuaded Samorn to walk about a third of the way to the wallow, but found she was determined to go no further. Patiently, over several more weeks, Andrew led her a little further each day until they finally reached the mud. Instantly, despondency gave way to excitement. After the wonderful discovery of an activity that had been a part of her regular ritual at the city zoo, nothing could keep her away from the wallow hole. The discovery had an immediate effect on her, enabling her to become more comfortable about her surroundings.

When Graham Goldsmith left Monarto, Paul O'Donoghue, who had worked at the city zoo some years before, came to fill the vacancy. On his first day at Monarto and against his better judgement (even though he had no choice), he had to go into Samorn's house on his own because there were only three staff. This was despite him being warned of her unpredictable behaviour, and her being untrustworthy and a beast that could do serious damage without trying. But Samorn, remembering him from years before, readily accepted him. When quizzed about her behaviour, he was quick to defend Samorn's reputation, recalling she was not as bad as others made out. He claimed she was not only a well-adjusted elephant but also a lot steadier than she had been at the city zoo where he first worked with her years before. After her transfer, when the staff at the city zoo were concerned about her settling in, several keepers took it in turns to visit her over a period of time, knowing how much she would welcome their familiar faces.

It was during her first month at Monarto that Samorn's most beloved former companion, Hero Nuus, died on 12 December 1991. Two days later, on behalf of Samorn, the zoo management placed her own tribute to Hero in the *Advertiser*. It read: 'Goodbye my special friend – Sammy the Elephant.'

Samorn with Andrew Mann at Monarto. News, *20 January 1992. (News Ltd – Newspix)*

After several months settling Samorn into her new environment, Andrew Mann decided it was time to leave Monarto. He explained that elephants are a little bit special because of the way you deal with them. 'Samorn was an important part of my life because I worked so closely with her.' He reiterated that there are very few other relationships like the ones you can have in a zoo situation.

It was six months before he returned to see how she was going. She immediately recognised him and began squeaking and rumbling, 'doing all the recognition things she'd do with the special people in her life'.

Andrew could only guess how naughty she would be the next day when she realised he wasn't going to be there, so he made an important but painful decision about future visits. So that her keepers would not have a difficult time handling her, it was almost three years before he went to see her again. As Andrew explained: 'I know how strongly I felt for she was an important part of my life. Well, when an animal is dependent on you for so much, imagine how they feel about you.' Andrew didn't see her again after the second visit. Ironically, Samorn died on his wife Donna's birthday in 1994. He sadly reflected, 'I know the day well.'

In 1991 it was deemed a good idea to relocate Samorn to Monarto. But in hindsight, one can consider she was used to her 35 years in the compact quarters of the city zoo where she could watch the passing parade of adults and children. She was used to the noises of the city, the sirens of ambulances, police cars and fire engines, and to the crowds and their chatter, and their continual interest in her. She was essentially a people's elephant who needed their constant interest.

Children have their favourites at the zoo, be it the giraffe with its extraordinary long neck, the proud lions and snarling tigers, the most venomous snakes in the world, the giant spiders or huge stick insects, the cheeky meerkats, or chimps and monkeys with their antics. For the youngest there is the Children's Zoo where they can be guaranteed an up-close-and-personal encounter or have their very first experience with tame animals. And there were always those who made a bee-line for Samorn, the zoo's largest animal. Fewer and fewer zoos have elephants, to the disappointment of many, and the star attractions are now often pandas, as at the Adelaide Zoo. But there are many who still believe: 'What is a zoo without an elephant?'

It is a wonderful vision to have a small herd of female Asian or African elephants at Monarto for there is plenty of room for a stampede if that's what a herd wanted to do. But the grim reality lies in the economics of supporting half a dozen elephants.

Although elephants can be obtained in exchange for other animals, costs such as their transportation to Adelaide would be exorbitant. Ed McAlister explained that to keep one elephant in 1991 cost around $95,000 per year, which also meant that 10,000 extra paying visitors to Monarto (paying ten dollars each at that time) would be needed for the upkeep of each elephant. This sum would finance the several keepers the elephants would need during the seven days a week (plus those to cover for holidays and days off), the extra food for a healthy diet, and the veterinary costs: in other words almost $600,000 a year for six elephants. Proud of his track record for keeping budgets firmly in the black, Ed admitted he held back the plans.

But he did expedite the opening of Monarto to the public after Samorn moved there and when the 15 kilometres of feral-proof fencing for the perimeter were completed. As he stated: 'It would cost us $200,000 to keep it shut or $200,000 to open it.' And so he convinced the zoo board to open it, at first on one day at the weekend, followed by both days of a weekend and holidays, before taking the plunge to seven days a week.

As Ed McAlister said, when you've seen one deer, you feel you've seen them all because they all look the same. The introduction of cheetahs and lions, the first carnivores, gave Monarto more focus. However, it was only opened to the public in October 1993 when Samorn took up residence at Monarto and the perimeter fencing was installed, after years of planning and empty promises.

Since then, Monarto has only looked to the future concentrating on its valuable role of contributing towards breeding programs of some of the animals, including endangered species. With attention to education and entertainment of families, it has become a premier tourist attraction in South Australia.

13

The end of an era?

SAMORN DIED IN OCTOBER 1994 at Monarto Open Range Zoo at the age of 44 years when, not long after eating her breakfast, she accidentally slipped into the moat of her paddock. She lay trapped upside down for about two hours before she was discovered. The moat was 90 degrees on the visitors' side and 30 degrees on the other, but because there was no rounded or scooped-out base she could not free herself from her precarious position.

On that sad day there were just three staff members and a few volunteers on duty. Paul O'Donoghue, one of the staff members, had earlier given Samorn her breakfast and gone off to do other chores when he was contacted about 10.30 am by two-way radio from the business centre of Monarto. The centre was alerted by the driver of the visitors' bus who, having spotted Samorn upside down in the moat while he was in the driver's seat, called them from his radio.

Alarmed, Paul drove to her paddock. It was some

time before he found her in the moat with her four legs in the air barely moving. He clambered down into the moat and put his hands on her head and around her ears to feel how hot she was. The young student who accompanied him was sent to fetch water from her wallow to pour over her to cool her down, for it was mid morning and already the day was heating up.

Having had years of experience with elephants, Paul was not hopeful of her surviving the ordeal but, nevertheless, he rushed off to seek help from the business centre. A crane was called from Murray Bridge to lift her out and two vets rushed to the scene to help. Miraculously, because it was stated to be an emergency, the crane was on site within half an hour. In that time Paul organised wide straps to fit around her body. In the 30 minutes it took for the crane to prepare for the task, a trench was dug under Samorn for the straps to pass around her body. These were then attached to the crane jib. She had not moved until she was being secured into the harness, but Paul could not avoid her legs when she suddenly began kicking violently. By the time the crane lifted her gently from the moat and she was righted with her legs tucked beneath her she had grown very calm.

But it was all too much for Samorn, exhausted from the ordeal of struggling to free herself in the period before she was found. Despite the hard work of lifting her out of the moat, Samorn, whom Paul described 'as a lovely girl, a nice steady girl', quietly died shortly after, not having had the energy to stand up.

The crane, while still at Monarto, was used to transfer Samorn's carcass for burial in a nearby quarry.

A tribute to Samorn by Ross Bateup following her death. Advertiser, *11 October 1994. (Newspix)*

When the day finally ended, Paul recalled: 'To be honest I was knackered.' He was not only exhausted and bitterly disappointed about Samorn's forlorn death, but bruised and sore from where she had kicked him.

A gloom descended on those working at Monarto and at the city zoo as news of Samorn's death was confirmed. It took a while before people who knew her recovered from the shock. Letters of condolence were received by the zoo and David Langdon, then the director of Monarto, went to the trouble of having cards printed and sent out to 'Thank you for your expression of concern'.

Eighteen months after Samorn's death, the University of Adelaide showed an interest in her carcass and she was exhumed so she could be retained for exhibition. Researchers spent six months assembling her vast skeleton for comparative anatomy and discovered she had suffered severe 'honeycomb' deterioration in one

The smaller sized skeleton of Samorn is alongside Mary Ann in the Museum of Medical Anatomy. (Photograph taken by the author, 2015)

toe bone indicating she had endured a very sore foot. On her death, the newspapers of the day stated she had been suffering from long-term osteoarthritis in 80 per cent of her joints.

Now you can see her skeleton in the foyer of the Anatomical and Medical Sciences Resource Centre in Frome Road, where she is positioned looking impressive alongside the larger Mary Ann, the zoo's second elephant. You could say Samorn at last has a long-term companion.

While Samorn never had a mate, she had Tara as her companion for several years, an animal in her own image that induced her to behave like an elephant with attitude. Despite Samorn's size and strength she was an adored celebrity in her own right for many years as the 'state's pet', rather than a wild animal or domestic pet. She is remembered affectionately by journalists, photographers, TV entertainers, poets, writers, cartoonists and, of course, all the grown-up children saddened by her death.

Importantly, questions need to be answered as to whether elephants have a future in our overpopulated world? The grim reality is that in Africa, India, Sri Lanka and Southeast Asia, their plight is nothing short of critical.

In the more reputable zoos around the world free contact by staff has mostly been replaced with 'protected contact'. Unfortunately there are far more zoos still expecting their animals to perform and entertain visitors. This includes the Bangkok Dusit Zoo where Samorn spent her last months in Thailand before travelling to Australia.

Clearly, there are plenty of concerned elephant aficionados around the world skeptical about elephant survival in the aforementioned countries. In Africa poaching for ivory to satisfy the desires of the growing numbers of middle-class Asians is having a catastrophic effect on elephant numbers, particularly in Tanzania and Kenya. In India, Sri Lanka and Southeast Asia human overpopulation, illegal logging and alienation of forests for plantations of tea and palm oil continue unabated, with tragic results for elephants.

Elephant organisations have been founded to raise vital funds to secure land for sanctuaries, such as the Assam elephants campaign 'Elephants on Parade', as mentioned at the beginning of this book. Because of the massive scale of the tragedy taking place within one generation, it seems that unless there is a groundswell of support from around the world, and from within the countries affected (as their citizens become more politically activate and outspoken), it will take a miracle to avert the catastrophe now looming.

In Sumatra, where 70 per cent of natural forest areas have been destroyed through illegal logging to create palm oil plantations, the remaining 2000 local wild elephants were 'officially listed as critically endangered' in 2012 because 'half of its population has been lost in one generation'.

Involvement with elephants in Sumatra began in January 2012 for the wildlife vet, Amber Gillett. She was invited to the Elephant Conservation Centre in Bengkulu to see Bona, a rescued baby elephant. Amber and two other former employees of the late Steve Irwin (who was also particularly interested in the plight of Sumatran elephants) were mobilised into action by the plight of this orphaned female baby elephant. Their involvement in elephant conservation was featured in 'The Chosen One' in 2013, part of the ABC's *Australian Story* series. It revealed the three friends 'are currently discussing how to redirect any surplus funds from the Bona campaign to help other wildlife'. More details about them can be found on the Save Bona website.

In Thailand there are several organisations directed to care for injured, infirm and homeless domesticated elephants. Some have been operating for several years, and most are locked into the tourism or ecotourism industry, providing activities for tourists to keep their sanctuaries viable. In the north of Thailand near

Lampang, the Elephant Conservation Centre puts on shows to raise funds for elephant care.

In the Mae Taeng valley, 60 kilometres from Chiang Mai in Thailand, Sangduen Chailert (better known as 'Lek'), approaches conservation differently. She runs the Elephant Nature Park, a semi-wild environment that provides sanctuary. Lek has won many awards for her efforts in rescuing elephants from abusive situations or from a lifetime of work. As far as possible, the rescued elephants are returned to the wild. (Her website is www.elephantnaturepark.org)

Antoinette van de Water began working as a volunteer for Lek at the Elephant Nature Park in 2002. She is the founder and driving force behind 'Bring the Elephant Home', a campaign that supports Lek's vision by working to create a better future for domesticated Thai elephants that are overworked, poorly fed, homeless or ill-treated. She has campaigned to have elephants removed from city streets through better education, collaboration with other elephant protection agencies, and demands for better legislation for the protection of Thai elephants. The organisation also raises funds to expand existing elephant habitats as well as reforestation campaigns.

The Great Elephant Escape, a book she wrote with Liesbeth Sluiter, tells the story of the elephants' plight. Find information about the charity and Antoinette's important work at www.bring-the-elephant-home.org. She reminds us that the Thai elephant population, 100,000 in 1900, was, at the time of her publication in 2006, between 2000 to 3000 domesticated elephants;

there are less than 2000 left in the wild. A sad plight indeed.

Thailand is the country of Samorn's birth. While I was there observing elephants and gathering information about them, the Thai people I spoke with (for they often wanted to know the reason for my visit) in government departments, hotels, or on pavements over a pad thai or a bowl of tom yum soup were surprised that an Australian was in Thailand because of an elephant from so long ago. They loved hearing that one of their elephants, sent to South Australia in 1956, was adored all her life and mourned when she died in 1994.

The very psyche of Thailand is entwined with its elephants. Its history features elephants in heroic battles, and the country was developed using thousands of working elephants. You cannot walk anywhere in Thailand without being reminded of an elephant. A local beer is called Chang (Thai for elephant) and labels on beer cans and bottles feature the animal. Virtually every temple carries elephant images in some form. They may be included in the architecture, or placed as statues around courtyards. Such statues range in size and materials and vary from those in humble plaster to extraordinary bejewelled ones, such is the people's reverence for them. They feature in murals on temple walls, and in advertisements throughout Thailand's cities and villages. And notably many tourists would leave Thailand with some reminder of an elephant. It might be a silk cushion cover, a carving, paper handmade from elephant droppings, bags, paintings or clothes. The irony is that elephant numbers continue to plummet.

It is unlikely that any more elephants will settle in South Australia. But we can ask questions about our last one, the revered Samorn, and where she came from. And as a tribute to her memory one must be very concerned for the plight of the remaining few left in Thailand – for they could all be gone within a generation.

References

Chapter 1

Mail, 21 May 1949 'Elephant pushed into moat'; *The Argus* (Melbourne), 24 July 1945, 'The story of Betty and Peggy'; C.F.H. Jenkins, *Noah's Ark Syndrome*, 1977, p. 138; Penny Olsen, *Glimpses of Paradise: the quest for the beautiful parakeet*, p. 231.

Chapter 2

State Library of South Australia (SLSA), PRG 50 Governor Gawler; Richard Aitken, *Seeds of Change*, p. 80; *Maitland Mercury & Hunter River General Advertiser,* 30 August 1851; *South Australian Register,* 4 April 1855.

Chapter 3

Advertiser, 10 May 1904, p. 4 'Death of the zoo elephant'; *Advertiser* 17 May 1904, p. 6 'The late Miss Siam'; Natalie Lloyd (Edwards), 'Little Worlds: Australian Zoological Gardens in Australia and their Imperial Connections'; Kay Anderson, 1995, 'Culture and Nature at the Adelaide Zoo: At the Frontiers of "Human" Geography' pp. 275–294; SLSA/SRG 263/7 – RZSA Council Minutes; *Barrier Miner,* 19 January 1932, 'Adelaide Zoo elephant attacks keeper'; *The Mail,* 20 January 1934, 'Zoo seeking new elephant'; *Advertiser,* 12 May 1937, 'Novel Liars' Club Contest Estimating Weight of Elephant'; *South Australian Register,* 15 August 1924, 'Bid for freedom – Circus Elephant Escaped'; *The Mail,* 14 May 1938, 'Carrots on Her Birthday Cake – Lillian the Zoo's baby elephant ...'; SLSA/SRG 263/42, December 1943; *Advertiser,* 8 September 1950, '"Mary" an elephant you will never forget'; *The Mail,* 7 October 1950, 'The world's first mechanical elephant'; *Sunday Times* (Perth) 22 October 1950, 'Toy Elephant is "Vehicle"'; *Advertiser,* 15 December 1925, 'An Adelaide Inventor'; *Advertiser,* 31 October 1933, 'Adelaide woman inventor'; SLSA/SRG 263/7 Council Minutes; Trevor Klein, verbal information, October 2012; Blair Csuti et al (ed.), *The elephant's foot: prevention and care of foot conditions in captive Asian and African elephants*; George Orwell 'Shooting an elephant'; *Australian Dictionary of Biography* for Sir J.K. Angas; J.K. Angas, *Safari*, 1922; *South Australian Register*, 15 July 1922, 'Big Game Hunting'; Trevor Klein verbal information September 2012.

Chapter 4

Ryhiner, Peter & Mannix, Daniel, *The Wildest Game,* pp. 19–26, 99–101, 215, 223; Ray Hoser, *Smuggled: The Underground Trade in Australia's Wildlife,* p. 23.

Chapter 5

Advertiser, 13 January 1940, p. 11 'The greatest elephant hunt in history'; National Archives of Thailand, Bangkok,

Ministry of Foreign Affairs, Code 71.1.1/4, 1954–1957 (2497–2500).

Chapter 6

SLSA/SRG 263/7 Council Minutes; *Advertiser,* 18, 20 October 1947; *Mercury (*Hobart), 8 July 1952; *Australian Women's Weekly,* 24 September 1980; Peter Ryhiner, *The Wildest Game*; Ray Hoser, *Smuggled*, p. 23; *Australian Dictionary of Biography* for Sir Edward Hallstrom; *Australian Dictionary of Biography* for David Cody; Steven J. Smith, *The Tasmanian Tiger*, p. 103; Adelaide Zoo Archives Newspaper clippings volumes for ABC 29 November 1956; *Advertiser*, 30 November 1956, 'Gift elephant to arrive today'.

Chapter 7

Advertiser, 30 November 1956, 3 December 1956; *Sunday Mail,* 29 December 1956, 'People from 7 to 70 all love her'; *Advertiser,* 6 April 1980, 'Manicuring Samorn's feet'; *The Mail*, 22 August 1959, 'Samorn used trunk as a paint spray'; *Advertiser,* 19 September 1961, 'Facing up with Fatchen'; SLSA/SRG 263, Annual Report 1973–74 for 'Twenty years the Zoo Vet'; SLSA/SRG 263/7 Council Minutes 1962; *Advertiser,* 6, 7 February 1962; *Border Watch,* 6, 8 February 1962; website for Sole Circus; website of Hilda Padgett, 'The hanging of Mary the elephant, Erwin, Tennessee', 1996; Lee M. Talbot, 'Ecological considerations in the use of wild animals for biomedical research', pp. 286–295; *Straits Times*, Singapore, October 1962; *Advertiser,* 22 May 1963, 28 August 1963.

Chapter 8

Advertiser, 20 October 1947; *SA Truth*, 18 January 1964; SLSA/SRG 263/7 Council Minutes for 26 August 1963, 'William Gasking's first impressions'; 'Bye bye Birdie' in *The Bulletin,* 4 July 1964; C.F.H. Jenkins, *The Noah's Ark Syndrome*, 1977, pp. 53, 131; *Advertiser*, 3–26 September 1963, 'Customs breaches alleged'; SLSA, SRG 263/7 Zoo Minute Book Vol. 9, 1957–1962; *SA Truth,* 18 January 1964; *Advertiser,* 14 November 1963; Stewart Cockburn, 'The Zoo Affair', *Advertiser*, 30 January 1982; Penny Olsen, 'Crooks, counterfeits and contraband' in *Glimpses of Paradise*, pp. 181–192; Libby Robin, *The Flight of the Emu*, pp. 202–207; Ryhiner, Peter & Mannix, Daniel, *The Wildest Game,* pp. 21–22; SLSA SRG 263/7 Minutes of the Adelaide Zoo, SRG 263/42 Director's Journal; *SA Truth,* 18 January 1964; *Advertiser,* 11 September 1963, 'Tara plays up'; *Advertiser,* 25 August 1954, 'Dolly the elephant on walkabouts'; Kay Anderson, 'Culture and Nature at the Adelaide Zoo' quoting Adams et al 1991; *Advertiser,* 17 October 1963, 'Visit by Joy Adamson'.

Chapter 9

SLSA SRG 263 Annual Report for 1973–74; *Advertiser*, 14 November 1963; SLSA SRG 263/58 'Confidential report'; *Advertiser*, 30 January 1982, 'The Zoo Affair', p. 24; *Advertiser*, 14 November 1963; verbal information by Leo Oosterweghel, 4 October 2013 about an animal known as Minchin; ADB for R.E., A., R.R. & A.K. Minchin; SLSA SRG 263/42 – Director's Journal, Vol. 10, 7 July 1935, 17 December 1937; C.E. Rix, *Royal Zoological Society of South Australia 1878–1978*, p. 49, 53; Raymond Hoser, 1993, *Smuggled: The Underground Trade in Australia's Wildlife*; *SA Truth*, 7 September 1963, 'Used zoo as sham – Crown alleges'; *Advertiser*, 8 January 1964, p. 2, 'Looking to the zoo's future'; *Advertiser*, 21 January 1964, 'Mr Basse steps down'; *Daily Telegraph* (Sydney) 2 December 1965; 'Who's who at Taronga', *The Bulletin* (Sydney), 13 August 1966; *Sydney Morning Herald*, 15 July 1966; C.F.H. Jenkins, *The Noah's Ark Syndrome*, 1977, p. 107; Ed McAlister, verbal information, November 2012.

Chapter 10
Advertiser, 16 December 1988, 'Elephant man was zoo favorite'; Leo Oosterweghel, verbal information, 2012; Ivor D. Balding, 'One lucky elephant' documentary; Trevor Klein, verbal information, Nov 2012; SLSA, SRG 263 – Zoo Minutes; *Advertiser*, 18 December 1990, 'Letter to the editor about Tara's death'.

Chapter 11
National Archives, 'Destination Australia', Hero Nuus & Bill Lancaster, 1/1965/16/9; Ed McAlister, verbal information, Dec 2013; Leo Oosterweghel, Dublin Zoo, verbal information, 2012; Werner Zur Eich, verbal information, June 2012; Victoria Raine, *Elephant in Our Zoo*, 1986; *Bendigo Advertiser*, 8 December 1986; Peter Whitehead, information, 2012; Andrew Mann, verbal information, April 2012; SLSA SRG 263 Zoo Minutes, 'Grand Prix', 1985.

Chapter 12
Advertiser, 16 December 1988, 'Hero farewells one giant affair'; *New Idea*, 5 May 1985; Andrew Mann, April

2012, C.F.H. Jenkins, *The Noah's Ark Syndrome,* 1977, p. 123; SLSA SRG 263 – Annual Report, 1991; Paul O'Donoghue, verbal information, Dublin Zoo, June 2013; Ed McAlister, verbal information, Dec 2012.

Chapter 13

Paul O'Donoghue, verbal information, Dublin Zoo, June 2013; Ed McAlister, verbal information, Dec 2012; National Archives of Thailand, Bangkok, Ministry of Foreign Affairs, Code 71.1.1/4, 1954–1957 (2497–2500); ABC *Australian Story*, 'The Chosen One', 2013.

Books and articles

Adams, G., et al 1991, *The role of the Adelaide Zoo in conservation*, Report prepared for the Royal Zoological Society of South Australia, Mawson Graduate Centre for Environmental Studies, The University of Adelaide.

Aitken, Richard 2006, *Seeds of Change: An Illustrated History of Adelaide Botanic Gardens*, Board of the Botanic Gardens and State Herbarium, Adelaide, South Australia.

Allin, Michael 1998, *Zarafa: the true story of a giraffe's journey from the plains of Africa to the heart of post-Napoleonic France.* Review, Great Britain.

Anderson, Kay 1995, 'Culture and Nature at the Adelaide Zoo: At the Frontiers of "Human" Geography' in *Transactions of the Institute of British Geographers,* pp. 275–294.

Angas, John Keith 1922, *Safari.*

Australian Dictionary of Biography for Sir J.K. Angas, David Cody, Sir Edward Hallstrom, Richard Ernest Minchin, Alfred Corker Minchin, Ronald Richard Minchin and Alfred Keith Minchin.

Bailey, Col 2013, *In the shadow of the Thylacine*, Five Mile Press, Scoresby, Victoria, Australia.

Beresford, Quentin & Bailey, Garry 1981, *Search for the Tasmanian Tiger,* Blubber Head Press, Hobart, Tasmania.

de Courcy, Catherine 1995, *The Zoo Story*, Claremont, Ringwood, Victoria.

de Courcy, Catherine 2009, *Dublin Zoo: An Illustrated History*, The Collins Press, Cork, Ireland.

Csuti, Blair et al (ed.) 2001, *The elephant's foot: prevention and care of foot conditions in captive Asian and African elephants*, Wiley-Blackwell, USA.

Guiler, Eric R. 1985, *Thylacine: The Tragedy of the Tasmanian Tiger*, Oxford University Press, Melbourne.

Helfer, Ralph 1997, *Modoc: The True Story of the Greatest Elephant that Ever Lived*, Harper Collins Publishers, Inc., New York.

Hoser, Raymond 1993, *Smuggled: The Underground Trade in Australia's Wildlife*, Apollo Books, Mosman, NSW.

Jenkins, Clee Francis Howard 1977, *The Noah's Ark Syndrome: 100 years of Acclimatization and Zoo Development in Australia,* South Perth, WA: Zoological Gardens Board of Western Australia.

Lloyd (Edwards), Natalie 2007, 'Little Worlds: Australian Zoological Gardens in Australia and their Imperial Connections' in P. Limb (ed.), *Orb and Sceptre: Studies in British Imperialism and its Legacies*, Monash University Press, Victoria.

Mazur, Nicole 2011, *After the Ark? Environmental policy making and the zoo*, Melbourne University Press.

May, Ken 2009, *Phaniat: Royal Elephant Kraal and Village Ayutthaya, Thailand.*

Olsen, Penny 2007, *Glimpses of Paradise: In search of the Paradise Bird,* National Library of Australia, Canberra.

Owen, David 2003, *Thylacine: The tragic tale of the Tasmanian Tiger,* Allen & Unwin, Sydney, NSW.

Paddle, Robert 2000, *The Last Tasmanian Tiger: The History and Extinction of the Thylacine,* Cambridge University Press.

Raine, Victoria 2011 (first published 1986), *Elephant in our Zoo*, Hyde Park Press, South Australia.

Rix, C.E. 1978, *Royal Society of South Australia 1878–1978*, Royal Zoological Society of South Australia, Adelaide, SA.

Robin, Libby 2001, *The Flight of the Emu: A Hundred Years of Australian Ornithology 1901–2001*, Melbourne University Press, Carlton South, Victoria.

Ryhiner, Peter & Mannix, Daniel P. 1959, *The Wildest Game,* Cassell, London.

Sluiter, Liesbeth and Van de Water, Antoinette 2006, 2009, *The Great Elephant Escape,* Silkworm Books, Gravenhage, The Netherlands.

Smith, Steven J. 1981, *A report on an investigation of the Tasmanian Tiger*, National Parks and Wildlife Service, Hobart, Tasmania.

Williams, J.H. 1956, *Elephant Bill*, Penguin Books.

Acknowledgements

After the publication of *The Adelaide Park Lands: A social history*, my publisher Michael Bollen of Wakefield Press casually asked what I was writing next. When I replied that it was a story about an elephant, his enthusiasm convinced me to pursue it. I give my thanks for his support.

One is always in debt to those who offer information tidbits, for such tantalising clues often lead to other sources. Several friends gave advice and suggestions, especially after they had a peep at my manuscript. They include Roger Andre, Hamish Angas, Geoff Bickley, Ann and Robert Bowman, Karin Dunsford, Brian Samuels, and Dr Natalie Edwards of Western Australia.

I thank the family of the late Max Fatchen (who I was in contact with at the time of his death) for the use of his poem. I am in the debt of Peter Goers of ABC Radio 891 for always supporting and promoting my contributions to South Australian history.

Cheekily, I sent my manuscript to the world's

leading elephant behavioural consultant, Alan Roocroft in Tennessee of Elephant Business, he who established 'protected contact training'. I am delighted for his enthusiastic support of my biography of Samorn.

Catherine de Courcy, writer in 2006 of the Dublin Zoo's history, gave me much encouragement and guidance about elephant behaviour. Thank you.

Among the zoo fraternity in Adelaide I thank past and current management and staff, and volunteers. These include Andrew Mann, former elephant keeper, Dr Ed McAlister AO, a former director of the Adelaide Zoo, Alison Murchie, zoo volunteer, Werner Zur Eich, former superintendent of mammals at Adelaide Zoo, and Dr Chris West, also a former director of Adelaide Zoo. Deservedly due, are many thanks to Theresa Case for valuable liaison between myself and the Adelaide Zoo.

I am in debt to Leo Oosterweghel, Director, Dublin Zoo, who as a former curator of birds at Adelaide Zoo inspired much enthusiasm for Samorn's story through many phone calls and my 2013 visit.

Paul O'Donoghue, another former Adelaide Zoo staff member now at Dublin Zoo (as assistant director), was with Samorn when she died. He told me of her last day while I was staying at Dublin Zoo.

I am most grateful to the present director of the Royal Zoological Society of South Australia, Elaine Bensted, whose support by a letter of introduction to the Thai National Archives in Bangkok was crucial for me to gain access to its collection. I am also in her debt for the use of several of the Society's photographs.

In Thailand I am in debt to Khun Patsri Tippayaprapai

(Mameaw) of Bangkok, who was my translator for two days. I also thank Khun Varanud Vinasandhi, the very helpful archivist at the National Archives in Bangkok.

I express my gratitude to Jaye Walton, OAM, and The Order of the White Elephant, who as the Honorary Consul-General for Thailand, South Australia and Northern Territory, made my journey to Thailand as a researcher less hazardous. I also extend many thanks to her charming assistant David Wright.

For the use of media archival images I thank Paul Jackson and Sharon Polkinghorne of News Limited who slickly guided me through their vast collection. For single images, appreciation goes to Peter Lindeman of Fairfax Syndication; Laura Smith of Archives of Appalachia, Johnson University, Western Tennessee; Penny Standen of the Elsa Conservation Trust, Sunbury-on-Thames, UK. Michel Angas, Yvonne Austin, niece of Douglass Baglin, Frank Manfield, Emma Simmons, Cynthia Taylor and Chris Thompson willingly allowed me to make use of their family photographs.

For having great faith in my view of Samorn's story, Trevor Klein, sales manager for Wakefield Press and an Adelaide Zoo volunteer, was a great source of zoo stories. He fed me an endless supply of valuable reading material and was my anchor through choppy seas. When the manuscript was completed editor Julia Beaven waved her magic wand and I felt so much better.

My biggest thanks are to my partner, Robert Martin, who not only bravely accompanied me to Thailand to seek out elephants and information about them but read and criticised the draft manuscript.

INDEX

P

Q

R

S

T

WHO AM I?

ANITA HEISS

An Omnibus book from Scholastic Australia

To my mother, Elsie, for her inspiration, strength and wisdom. And to all those who are still asking 'why'? May you find peace, even if never an answer. — A.H.

LEXILE™ 1000

Omnibus Books
An imprint of Scholastic Australia Pty Limited (ABN 11 000 614 577)
PO Box 579 Gosford NSW 2250
www.scholastic.com.au

Part of the Scholastic Group
Sydney • Auckland • New York • Toronto • London • Mexico City • New Delhi • Hong Kong • Buenos Aires • Puerto Rico

First published by Scholastic Australia in 2001.
This edition published in 2021.

Cover design by Chad Mitchell.

NATIONAL LIBRARY OF AUSTRALIA

A catalogue record for this book is available from the National Library of Australia

ISBN: 978-1-74276-958-5

Typeset in Berkeley Oldstyle.

Printed in China by Hang Tai Printing Company Limited.

FSC www.fsc.org MIX Paper | Supporting responsible forestry FSC® C023121

10 9 8 7 6 5 4 3 26 27 28 / 2

The Diary of Mary Talence

Sydney, 1937

Bomaderry Aboriginal Children's Home

Wednesday, 27th January 1937

My name is Mary Talence. I used to be called Amy Charles, but they changed my name when I came to live here at the Home five years ago. I was only five then and today it's my 10th birthday. And you were a birthday present from Matron Rose. She's in charge here and almost like a proper mum to me. She always gives me a birthday present cos I'm her favourite. She likes me so much that she let me stay here at the Home much longer than the other kids. Most of the girls that come here have to leave for Cootamundra Girls' Home by the time they are ten years old, and the boys have to leave for Kinchela Boys' Home. That's the really dark kids though. The light-skinned ones, they go to other orphanages like the Burnside homes at Parramatta, Lutanda or Wentworth Falls. Those homes are for all kids without parents, not just Aboriginal kids. But not me, they let me stay longer and I help with the little kids. I feel like this is my home and I don't want to have to make a new family at Burnside or 'Coota'—that's what they call the girls' home. Some of the other kids even go and live with new families. It's funny cos the ones that go and live with the new families are not the dark kids but the light. It's meant to be easier for them to fit into white families and schools and stuff like that.

I've got light-brown skin and Matron calls me her 'coffee-coloured princess'. I've got long dark brown curly hair and brown eyes. There's kids here though that are really, really dark brown. They must have got lots of sun I reckon. We sometimes have to line up from the darkest to the lightest, so they know which ones have to go to which homes, and who can be fostered to white families. None of us kids really know what's goin' on at all. We just do as we're told cos the Matron is like a real mum so she must know what's best for us, eh?

My best friend is Marj and she's my age. Matron calls her the 'chocolate princess' cos she is really dark with big brown eyes and dark brown hair. She laughs real funny and loud. She sleeps right next to me and we are the bestest friends in the world. We tell each other everythin'. I think we'll be friends forever.

Matron Rose said I should write in you every day about all the things I do and stuff. And I do lots of things that I'm gunna tell you about. So, I'm gunna make you my new best friend, and tell you secrets. But I have to go to bed now so I'll write again tomorrow.

Thursday, 28th January

So I guess you wanna know how I got here. Well, I don't remember a lot, cos I was only about five when I was brought here with my two baby brothers Bluey and Busta, my sister Betty and my cousin Max. And the day we left our real family was the saddest day of my whole life. My aunties and my mum was all cryin' and Mum was sayin' she didn't want us to go, but the men in the suits just said we had to go with them. My baby brothers and sister was so little they don't even remember but they was cryin' cos they was scared, I think. I still don't know why we was brought here.

Bluey, Busta, Max and Betty are all gone now to the other homes and I haven't heard from them. I miss them a lot and cried for days and days when they left. I thought my heart had broken into lots of little pieces cos now my real parents and my brothers and sisters was all gone from my life. I wanted to go too cos I wanted to stay with my family, but Matron Rose said it would be better if I stayed with her at the Home. I think about them all the time. I hope they're all right.

I also have two older sisters, Margaret and Janie, but I don't know where they are, or my older brother, James. Maybe they're still with our real mum and dad in the country. I asked Matron Rose about them a few times, but then she said I wasn't allowed to talk about them no more, so I stopped askin'. Marj is kinda like a sister now.

The day we got here we didn't know what was goin' on. I thought we was goin' on a holiday or somethin' cos we had this big trip on the train. I was a bit excited cos we got sandwiches and biscuits. But Mum and Dad weren't with us so I didn't really know what was goin' on. I just kept lookin' after the little ones.

All the Sisters here are nice, but no-one ever told us why we was here or why we didn't live with our parents no more. Mum came to visit a couple of times and then she didn't come no more. Maybe it's cos she was always so sad. Or, maybe she didn't want to see me no more. I wonder if I did somethin' bad? She cried a lot and it made me cry cos I missed her so much. I miss her huggin' me and singin' to me. She taught me to sing good. I love singin' and now I sing all the time too, so I'm really like her. When I get sad now though, Matron Rose always gives me a hug, and she says I should write to you about what I feel, which is what I'm doin', eh?

I'm lucky I can remember my real mum and dad. My mum was really beautiful with long black wavy hair right down to her backside. She wasn't really tall, not as tall as my dad anyway. Dad was really tall. Some people called him 'Stretch'. He had really short hair and green eyes and was really muscly—musta been from carryin' wood to build things. He was always buildin' things. That's about all I can remember my dad doin'. Oh, he liked to fish too. He was a good fisherman, and my mum was a good cook, always makin' big stews for the whole family to eat. I miss Mum and Dad a lot and think about 'em nearly every day. But I don't talk about them. No-one talks about their parents or their families. There's just a big silence around that. Even when I hear some of the little girls cry at night or have nightmares or when they wee the bed all the time, I know it's cos they miss their mums but no-one ever says anything. Not even Marj or me. You just don't.

But lots of the little kids don't even remember their mums and dads anyway, cos they came here when they was babies. They even think the Sisters or Matron are their mums, so they call Matron Rose 'Mum'. She doesn't mind. I think she likes it. And they call the other workers Aunty so-and-so. It's kinda like a family here, a really big family, eh?

They even brought a new baby in today. She's only a week old. That's really small. Some of the babies that come here are really, really tiny, like little brown dolls. So this is the only family they know. Other babies are a couple of months old, but they are still too small to know who their mothers and fathers are. I wonder where their parents are, and why they came to live here, like I had to. I asked Matron once why kids come here, and where their parents are. She just says 'It's for the best,' or 'Their parents are dead'. She doesn't say how they died though, so that's really strange. I wonder why they all died. I didn't ask though cos no-one talks about that stuff.

Cos I'm really good helping with the little children, and some of them are babies, I was told I was real grown-up by the Sisters here. I think that's why they let me stay, so I could help like a grown-up.

Friday, 29th January

Wanna know what I'm doin' when you're under my pillow? Well, like I said, I'm the oldest as most of the kids are babies and young school kids, so I'm like everyone's big sister. I really like that. Maybe that's why Mum let me come here, cos I was good at home helping with Bluey, Busta, Max and Betty, and she thought I should help the Sisters with these kids, too. But there's lots more to look after here.

Anyway, there's about forty kids here at the moment, about twenty boys and twenty girls and most of us all get on like one big family. Matron says that forty-seven kids is the most that can be here at any one time. There's about six Sisters too, one Brother, called Brother Andrew but we call him Uncle Andrew, plus Matron Rose. That's a big family, eh? And that means lots of dishes to wash. I reckon that after dinnertime would be a good time to be a book cos you don't have to do the dishes! It's not that bad though, cos Marj and I do them together and have fun.

Saturday, 30th January

Saturday is my favourite day cos we do fun things like going to the beach and bushwalking. Bomaderry is on the Shoalhaven River near Nowra so there's lots of nice countryside around to explore. Today we went to the Bomaderry creek for a swim. The Sisters know I love to swim so they didn't make me watch the little kids at all, so I could just swim and play with Marj. We climbed trees too and little Dusty got scared and wouldn't come down till one of the Sisters came to help. I got in a bit of trouble for letting him climb so high. He's only four, you see. See, I'm still meant to be the big sister all the time.

But I like Dusty cos he reminds me of my cousin Max who used to climb trees and then for attention pretend he couldn't get down. I really miss playing with Max and my other cousins. I wonder what they are doing, and if they are with my Aunty Barbara or in a home like this one.

Sunday, 31st January

We went to church today cos you have to be VERY Christian here. We go to the Baptist Church in Nowra and we say prayers with the Baptist Missionary there and have a sermon. Sometimes it's interesting, but sometimes it's boring. See, all the Sisters here, and Brother Andrew (he fixes things around here), they trained at the Baptist Missionary College cos they wanted to be missionaries with Aborigines.

Anyway, it was really hot today so Brother Andrew took some of the little kids in his car, and the grocer man with the big truck let some of the kids sit in the back with the fruit. But the rest of us had to walk there and back which is a pain. At least we had our good shoes on and our Sunday best clothes. For the girls that means really good pleated tunics and colourful head bands. Mine's blue and Marj's is red. We also wear knee-high woollen socks and shiny black patent-leather shoes. The shoes are a really big deal cos we don't wear shoes during the week at all. All our Sunday clothes are locked up in a cupboard at the Home so we don't wear them during the week. These nice clothes are donated to us from Baptist churches around Sydney. There must be some really nice people in Sydney, eh?

Cos we're Christians we pray a lot here at the Home. We say prayers before we eat meals to thank God for the food we are about to eat. We also pray after meals and then sing about three or four traditional Baptist hymns. That's like praying to music, you know.

With all the praying I do I am sure to go to heaven. Heaven is in the sky above the clouds and it's where everybody's souls go to when they die. But only people who are good when they are alive go to heaven. Bad people go to hell where it's really hot and the Devil is in charge down there. It's below the earth somewhere.

Monday, 1st February

I went to school today, just like every other day (except for Saturday and Sunday of course). At 7 am every day we have our breakfast of porridge or bread and treacle or bread and jam, and then we walk two miles to Bomaderry Primary School. We have about half an hour to play before we start school at 9 am. All the kids from five years and up have to go to school. It doesn't matter how dark or light you are, you all have to go to school. None of us kids have shoes and it's really bad in winter, cos it's freezing. Sometimes there's even ice on the ground and we pick it up and eat it. We end up with cracked feet from walkin' in the cold. The Sisters make these really big tubs of soapy water with lard and other stuff to rub into our feet to make them soft again. I reckon we get that frost-bite thing that people talk about. But it's hot now so it's not so bad, though the ground sometimes gets really hot.

We take sandwiches to school wrapped in brown paper or sometimes they're just all put together in one big sugar-bag. I don't mind school. We get to learn how to read and write and do sums and stuff. The only time I don't like it is when some of the kids call us 'home kids'. They make it sound like it's a really bad thing to live in a home, but they don't know that some of these kids' parents are dead or whatever. Marj got really upset when someone did that today.

Tuesday, 2nd February

Well you're still getting to know what my life is like, eh? Much more exciting than being a book, I reckon. But it's much harder work than just lying under my pillow or in a drawer all day too, cos here at the Home we have chores to do. Some of us help in the nursery with the little babies, feeding them or changin' their nappies. Other kids have to work in the laundry or wash dishes. I like doing the kitchen duty in winter cos there's a big stove and it's really warm in the kitchen when the Sisters are cooking. We don't have to help cook though. But someone has to clean the grease trap and that's hard work, throwing hot soapy water or caustic soda in it.

If we don't get our chores done or if we don't do 'em properly then we get in trouble and get hit on the legs with a stick or a switch from the tree. We only get hit on the arms or legs though, never on the head or anything like that. I've never had it though cos I always do my chores properly, but the boys are always gettin' in trouble.

I don't mind doin' my chores. I just sing all the time and so does Marj, but I'm a bit better than her. She can't hit the high notes like I can.

Wednesday, 3rd February

I need to tell you what the Home is really like so you know where you're living too, eh? Well, the Home is about thirty years old now and it used to be a cottage with four rooms and a kitchen on two-and-a-half acres of land. But they've added more cottages cos they needed them for all the kids. There's the nursery cottage for all the little babies. They stay there until they're about three years old. There's another cottage with two dining rooms (the children's dining room and the Sisters' dining room), the kitchen, and the pantry. There's the Sisters' cottage too where they sleep. There's also verandahs right around the cottages. All the kids from seven to ten years old sleep on the verandah. There's also a middle-dorm for the kids from four years to seven years to sleep in. The boys and girls have separate places to sleep and separate bathrooms with really big baths that can fit four girls in at one time. Bath time is lots of laughing and splashing. I wanted to sleep outside last night but the mozzies get so bad in summer cos the windows are open and they come right in and eat us alive. When I showed Matron my bites this morning she said I must have sweet blood. There's also a living room where all the officials can sit when they come to the Home. Us kids hardly ever go in there though.

And you know, Matron told me the Home nearly got burnt down by a bushfire once.

Thursday, 4th February

It's funny cos, before I got you, I'd be playing or looking after the little kids but now I really look forward to writing in you every day. It's like talking to a friend. I used to write letters to my mum and dad, just short bits to say hello and to tell them that I'm okay and that I love them. The sisters would read them and they said they'd send them but I never got a letter back so maybe Mum and Dad never got them. I don't know. So I stopped writin' a while ago. Marj never gets no letters neither.

Friday, 5th February

Every day Marj and I do the younger girls' hair for school. We give them really pretty curls and put ribbons in their hair and sing to them while we do it. They always look really cute in their pinafores and curls, even if they don't have shoes.

Today when I was singing it made me think of my real mum. It made me sad cos I miss her so much and wonder why she hasn't come to visit for so long. I hope she is still alive. The kids must think their parents don't come to visit us cos they don't love them any more, but how can that be true? Every parent loves their kids, just like how kids love their mums and dads forever. I think about my real mum every day and I'll always love her, no matter what.

Saturday, 6th February

It's the weekend again. Yippee!! Today we went flower-picking around the Home. We got some lovely flowers like flannel and waratahs and some of the Sisters got purple irises and yellow and pink waxflowers. The Home looks really bright now with all the flowers around the place.

We also picked lilly pillies, blackberries, bush cherries and wild cherries. Marj and I ate them straightaway and had really dark purple mouths all afternoon. It looked real funny and Marj pretended it was lipstick. When I grow up I want to wear real red lipstick like some of the ladies I've seen in the street. Matron says only certain kinds of ladies wear bright red lipstick and nail varnish.

Monday, 8th February

I'm a fast runner—everyone knows that—and there's going to be a big sports day at the school next week. I wanna go in it, but everyone has proper sports clothes to wear except for me. When I told Matron about the sports day I got a bit upset, but she said not to worry, she'd make me something to wear. I'm so happy, cos I know I can run the fastest, and I'll look even better doin' it in my new sports clothes.

I didn't eat too much bread and dripping for dinner cos I don't want to be too heavy to run next week. I hope we have soup and broth for the next few days, and no custard for me, that's for sure.

Tuesday, 9th February

When I was doing the girls' hair for school today I was just singin' silly words and I made a new song up for them to sing to Matron. They sung it to her this afternoon after school. I think she really liked it. It goes like this:

> Matron is the best mum
> The bestest in the world.
> Matron is the kindest lady,
> And the prettiest girl.
> Matron reads us books out loud
> And keeps us clean and well.

Matron is the best
And lets us ring the bell.

Wednesday, 10th February

The kids was singing Matron's song all day today and she couldn't stop smiling. She said I'm her favourite little coffee-coloured princess and Marj is her chocolate-girl, which I thought sounded funny. Imagine if she really was chocolate. She'd be licking her fingers all the time.

Thursday, 11th February

As usual I took the Sisters' bike to get the mail and to pick up some eggs from the shop near the home. I always get the mail cos I'm the oldest and Matron trusts me with the money not to buy lollies. I get the mail which is in a big leather bag with a lock with 'Aborigines Mission Home' written on it. Then I always go for a bit of a ride for fun.

The shop is nearly as far as the school and past the football oval. There was some boys playing footy today but I wasn't interested. There was lots of screamin' and cheerin' and whistle-blowin', and the players was all covered in mud cos it had been rainin' all night.

I hate being dirty now, but didn't mind when I was

a little kid in the bush and we'd play all the time with no shoes. We didn't have any to put on anyway. Now that I'm more grown-up I like to keep clean and act like a proper girl. Dusty loves the mud though and he was real jealous after I told him about the footy oval bein' all muddy.

Friday, 12th February

I'm so sad today cos some of the girls left for that girls' home at Cootamundra. Marj too. I can't believe it. The men in suits from the Protection Board and that woman came to get them and take them on the 7 am train to Sydney. Six went: Beryl, Marj, Ivy, Joan, Iris and Joy. We all cried cos it was like saying goodbye to our sisters or somethin'. I wanted to go with them cos they are my friends, especially Marj, and we always play and laugh together. But Matron said I was too fair to go and that I had to stay and help her with all the babies cos some new ones was coming in a few days. But I'm so sad cos Marj, she was my bestest friend in the world, even before you, cos you're a book and not a real person. My nose is all red from crying. I look all puffy like a red balloon.

Saturday, 13th February

We had the church picnic today at Husskisson. We did lots of swimmin' and played games but it wasn't the same without Marj there. I usually like the church picnic and wait for it to come round twice a year cos all the kids have a really good time. We all love the bus ride to Husskisson and I always nurse the babies on the way back. I thought about Marj all day when we was there, cos we used to pretend we was mermaids in the water. I wonder what she's doin' right now in that other home.

Sunday, 14th February

In church today I prayed for Marj and all the other girls that went to Coota this week and asked God to take care of them and bring them home to us one day soon cos I miss them all, especially Marj.

It's just not the same with the girls gone, especially Marj cos she really was my best friend. I cried myself to sleep last night cos it's like all the people I care about are always leavin' or goin' somewhere and then I just stay here and help Matron and that's it. I'm gonna write a letter to Marj and find out what it's like where she is. I'll go do that now and ask Matron to post it tomorrow. I already prayed that Marj would get my letter real quick and write me back.

Monday, 15th February

I wrote that letter to Marj today. It goes like this:

> To my bestest friend in the world, Marj,
>
> It's just Mary writin' to you even though you only left a couple of days ago. I miss you so much, playin' and singin' with me and everythin' like that. The other kids miss you too, don't worry about that. And how are all the other girls? And how are you? And what's it like at Coota? Please write me back and tell me what it's like there. Is it like Bomaderry at all? What do you have to do there? And are the Matrons nice? Oh, we went on the church picnic yesterday and I didn't even pretend I was a mermaid cos you wasn't here.
>
> Lots of hugs and kisses from your bestest friend,
>
> Mary Talence xoxox

Tuesday, 16th February

You won't believe this but the kids are all singin' Matron's song all the time and so I had to write a special song just for the other sisters cos I think they might have been a bit jelly bean (that means jealous) or something. So I wrote this song really quickly and taught the kids so they could start singing it straightaway and stop the Sisters being so jelly bean of Matron. It goes like this:

Sisters Sisters
So nice and happy
Sisters Sisters
Please change my nappy.

Sisters Sisters
So friendly and true
Sisters Sisters
We love you

Wednesday, 17th February

Seems like that song worked as all the Sisters are smilin' again and friends with Matron, thank goodness. They are not supposed to be jelly bean of each other cos jealousy is a bad thing—like a sin the Minister said in church one day.

I wonder how Marj is.

Thursday, 18th February

We had the sports carnival today and when I put on the shorts Matron made me they was all wrong. They looked like a skirt. She called them culottes. I wanted to cry cos I didn't want to be different and I told her that all the girls were wearing shorts to the sports day. She says, 'No Baptist girl is going to dress like a boy.' So I wore them to school and, you wouldn't believe it, all the other girls wanted a pair as well. They thought I was really good havin' a pair of shorts that look like a skirt.

I ended up comin' second in my race but I didn't care cos I looked so good and all the girls wanted to talk to me. I think Matron is really smart. She musta known what was gunna happen. Not only that, she made me a new blouse to go with the culottes and made sure I had proper shoes and socks too. I really am her favourite I think, cos the other kids don't get as much as I do.

Friday, 19th February

I'm not going to write much cos I'm really tired tonight. Matron has just left and said it's lights out at 8 pm. It's already 7.30 pm and she just finished telling us a story from the Bible. She tells us stories nearly every night and we all love them, and they're nearly always from the Bible. One day she told us a story and it wasn't from the Bible and we couldn't believe it. We're just so used to the Old Testament and stuff like that. It helps the younger ones go to sleep too, which is good. I'm yawnin' now, so it seems like the stories help me go to sleep too. I'll write ya tomorrow.

Saturday, 20th February

We had fun today just playing around the Home. Some of the kids played marbles and jacks, and some of the bigger girls skipped and played on the swings. I wanted to play vigoro which is like girls' cricket or rounders, cos I like team sports, but Matron said we could play next week. She also took me aside and told me some new kids are comin' this week, some little babies too, so we're getting ready for them.

I still haven't heard from Marj. I hope she hasn't got another best friend.

Sunday, 21st February

Because I wrote them two songs for Matron and the other Sisters I was asked to be like a conductor in church and get all the kids to sing today. I was really nervous but I liked being able to do that cos I really love singin'.

Matron says I have the prettiest voice of all the girls she's ever known in the Home. Today we sung that same chorus we always sing: 'Hallelujah, He's my friend, what if sorrow comes tomorrow, Hallelujah, He's my friend.' There's not many words but some of the little ones still didn't know the words proper and it sounded not too good at times. I might make them practise during the week for next Sunday's service.

Monday, 22nd February

Them new babies arrived today and I've already picked my favourite. He didn't even have a name. Matron Rose said I could pick one so I called him little John, like from the Bible. He was one of Jesus's helpers. And little John can grow up to be my helper. He's so little and cute, his mum and dad must be really good-looking people. I wonder where they are and why he is living with us. I asked Matron and she just said he didn't have any family and that we'll be the best family anyway.

Tuesday, 23rd February

I made the little kids practise some hymns today for church on Sunday so we don't sound so stupid in front of the Minister. We spent almost an hour just trying to get 'Amazing Grace' right. I think I'll have to make them practise again before Sunday. It's a hard job but someone has to do it. Matron really doesn't have a good voice and she's too busy anyhow.

Wednesday, 24th February

I am just about to go to bed, with the mozzies of course, and just wanted to say I had a nice day today. We've just finished telling each other scary stories. I don't know why we do it but we do. Then we giggle ourselves to sleep. Oops, there's Matron's knock on the wall, and that means lights out. It must be about 8 pm or maybe a bit later, and it's not even dark. In winter we have to be in bed by 7 pm but that's okay cos it's so cold and dark you just wanna get under the blankets anyway.

Thursday, 25th February

I've been singing a lot this week and it made me remember when I was a really little girl and my mum would sing to me all the time. She sang this one song that I'll always remember, but somehow I can't remember the last couple of lines:

Beautiful, beautiful brown eyes

Beautiful, beautiful brown eyes
Beautiful, beautiful brown eyes
I'll never . . .

Oh, I wish I could remember the rest of it.

Friday, 26th February

I made the kids practise again today. It's so hard to keep them interested cos they just want to play all the time. We sang the same hymn about ten times and then I couldn't do it any more so I let them go. I hope they are better on Sunday at church.

Saturday, 27th February

Still practisin' with them kids. They need a lot of work.

I wonder when Marj is goin' to write to me.

One of the new girls is real nice though. Her name is Catherine. She's real light, almost like a white girl.

Sunday, 28th February

The little kids were a bit better in church today cos we had that practice during the week. Dusty started laughing though cos I got angry with some of the boys who wouldn't sing loud enough. That made me more mad cos he was makin' fun of me. I'm the eldest and he should take more notice of me. But I'm not the Matron so I can't yell at him or nothin' like that.

Monday, 1st March

I am so sad today cos some of the boys left for the big boys' home called Kinchela. They took Catherine too, to go live with a white family. The Protection Board people came and took them on the 7 am train from Bomaderry to Central Railway in Sydney. Then they go to Kinchela from there. There was the same men and a woman who came for the girls last time. It's nearly always the same men in suits, but sometimes Matron or one of the Sisters goes with them to Sydney. Matron says Kinchela's a long way away and that boys go there to learn how to do manual work. I'm not sure what manual work is, but I think they have to scrub floors and wash things.

I heard Matron say to one of the Sisters that the manager man over at Kinchela used to drink a lot of alcohol and that some of the boys weren't safe there. You see, alcohol makes you drunk and do stupid things. Baptists don't drink alcohol. I don't think they smoke cigarettes either. Matron says we should never drink alcohol cos it's the Devil's drink. I don't know why the

Protection Board would send the boys to that place if they aren't gunna be safe. Matron said they have to do two hours' work every day after school in the dairy or the vegetable garden. That sounds like a lot to me. They must be so tired by the time they go to bed at night, eh? I wouldn't be able to write to you every night if I had to do all that work like them boys. I hope we'll all see each other again some day. That would be nice, eh, when we're all grown-up and everything to see each other again? It's good Catherine's goin' to a white family, eh?

Tuesday, 2nd March

Dusty got in trouble today for being cheeky to Matron Rose but it was just cos he missed the boys so much. All his big brothers from the Home have gone now and he's lonely. It's not like the girls are like boys or somethin'. He went climbing on the big coral tree and the weeping willow down near the creek and just sat there for ages. I reckon he must be really sad, but I don't know what to do about it.

Wednesday, 3rd March

Dusty was a bit betta today cos Matron told him he'd be going to Kinchela soon enough, and then he could see the boys. But she didn't tell him about that drunken manager man over there that is mean to the boys and makes them work too hard.

Thursday, 4th March

I can't believe it! Matron Rose told me today that I'm leaving Bomaderry. I just can't believe it! It has been a bad week with Catherine and the boys goin' and now me. I don't wanna go, but she says a nice white family in Sydney want a daughter just like me—all grown-up and sensible. She reckons that I'll be really happy there, and I can't stay at Bomaderry any more cos I'm too old and I'm not dark enough to go to Coota. Only the dark kids have to become domestic servants. I don't even know what a domestic servant is or does, but it doesn't sound too good. I think it's like a slave. But Matron thinks I should go with this family cos they have a lot of money and can send me to a good school and I'll have nice things. She said I might even get a guitar to help with my singin'. She knows how much I like to sing so that was a good way to make me think better about goin'. I think I'm leaving on Saturday. I'll let you know.

Friday, 5th March

Matron Rose is taking me herself on the train to Sydney to meet the family at Central Railway. She says all the paperwork is done and it's cos I'm nearly white that I can stay with this family and have a good life. She gave me a Bible to take with me so I can pray when I get scared or lonely, but I'm taking you with me cos I'll have no friends at all. Uncle Andrew gave me some really good pencils so I can draw pictures or colour in when I get bored on the train. I really love the hankies that the Sisters gave me that are all lacy and really white. They're not the sort of hankies you put snot on though, so I'll have to use my old hankies when I need to blow my nose and stuff.

So, dear diary, you'll be my only friend when I leave Bomaderry, but Matron said I can write back to her all the time and she'll write to me with all the news of the little kids and especially little John.

St Ives

Saturday, 6th March

I left Bomaderry today and everyone cried, the kids, the Sisters, even Uncle Andrew and especially me.

It took us a long time on the train, about five hours from Bomaderry to Central Railway where we met the new family who were waiting for me. I saw Matron crying as we drove off, but if she was so sad about me goin', why did she let the Protection Board send me away then?

I just sat in the car and looked out the window all the time and didn't say nothin' to the adults in the front of the car who are my new parents. My new mum smelled real fancy and had lipstick on. I wonder what Matron would have thought about that. I fell asleep for a little while too which helped, except I had a bad dream. I tried really hard not to cry. Matron told me not to, but it was really hard cos I'm going to live with these people and I don't even know them. I don't know anyone here in Sydney.

I'm at my new place in Sydney now, a place called St Ives. I'm gunna live here with my new parents, Mr and Mrs Burke, and their kids, Sam and Sophie. They said I could call them Mum and Dad but I feel strange about that cos I have a mum and dad. Even if I don't talk about them to anyone, it doesn't mean I don't remember or love my real parents. And I have Matron Rose who's been my pretend mum for the last five years. I think I'm gunna call the Burkes Ma and Pa B, if they don't mind.

I'm sittin' on the toilet writing this. I don't want to sit in my bedroom, cos it's not my room. It's my new sister Sophie's room and I know she doesn't like me. She told me already that she didn't want a sister, and especially she didn't want a sister that wasn't white. She says I'm brown and she's white. But she's not really white, more like pink with lots of freckles, and straight blonde hair that she has in pigtails with floppy ribbons.

Sunday, 7th March

I woke up this morning and I couldn't stop crying cos this place is not my home, even though everyone says it is. I miss Matron Rose and all the kids and now I miss my real mum more than ever. When I was a little girl Mum would always hug me when I cried and tell me everything would be all right. Who's gunna hug me here? Not Sophie, she hates me. She kept huffin' and puffin' all night cos I was in her room and she kept tellin' me not to touch her things. I don't even wanna touch her things. I have my own things, but she has more. She has lots of pretty dresses. I've never seen so many dresses—and lots of dolls. She is very spoilt, I think. But Ma B said she'd take me shopping this week for some new clothes for school and Sunday best.

Oh, the Burkes must be really religious too cos we had to go to church this mornin' and it was really strange as I was the only person there who wasn't white. I felt like everyone was lookin' at me. Ma B held my hand and introduced me to everyone, and they was all sayin' how good a person she was for taking a coloured child in. I hadn't heard that before, that I was a coloured child. Maybe that's another way of sayin' Aboriginal.

Monday, 8th March

I had my first day of school today and it was horrible. The school isn't far from where we live so I can walk to it. Pa B takes the car and drives to where the train station is and then gets the train into town. He's an accountant with Perpetual Trustee Accountants in the city. He smells very pretty too, like Ma B. He has thick black glasses and his eyes are big and scary through them. Anyway, Ma B walked with me and Sophie to the school. It's St Ives Public School and the schoolmaster, Mr Smith, lives in a house next to it. Ma B says she'd rather we all went to a Catholic school but there's not one in St Ives and it's too far to go to Lindfield or Pymble by ourselves. Maybe when we are in secondary school, she said, cos we can get the bus there.

Ma B took me to meet Mr Smith and my teacher, Miss Foster. They seemed really nice, but the kids on the playground at lunchtime were really mean to me, calling me all kinds of names and stuff. Like the lady at church yesterday, they said I was coloured. I said, 'No I'm not!' and they laughed at me. And now these kids was saying I was an 'abo' and made out that I was one of the bad ones. The worst kid was a really chubby boy called Johnny Jones who spat at me and called me 'boong'. I said I wasn't a 'boong', he was a 'boong', but I don't even know what 'boong' means. He seems like a really bad person, but he had lots of friends around him laughin' and agreein' with him. There was no other kids who looked brown at school so I had no-one else to talk to about it, so I went to the toilets and cried until it was time to go back into class. I don't want to go back tomorrow. I hate that school and Johnny Jones and I hate Matron Rose for letting them send me here. I wonder if Catherine has a nicer family. I hope so.

Tuesday, 9th March

I've only been here a couple of days and I hate it. I'm gunna run away back to Matron Rose. But what if she doesn't want me there? She mustn't want me there cos she let me leave with the Burkes to live here in the Hives (that's what I call St Ives). Matron cried when I left though, so maybe that means she's really gunna miss me and will be happy if I run away back to her. Oh no, Ma B is calling for me. I have to wipe my tears cos I know they don't want a sooky girl. They thought I was a real grown-up. That's why they wanted me. And Matron said not to give them any trouble, so that means not acting like a baby or nothin' like that.

Wednesday, 10th March

School was the same today. No-one talked to me—not even Sophie. She's younger than me, only nine, and her and all her friends play dolls at lunchtime and skip around the playground. Sam is twelve and in Grade 6, and was with all his friends and they look like the school bullies. He scares me a bit and hasn't even talked to me since I got here. I just sat on the playground by myself. I wanted to cry but didn't cos I don't want people to think I'm a crybaby. I didn't see any other brown kids either. I wonder where they all are. Maybe they are all on holidays or something. Maybe they are all at another school in the Hives.

I asked Ma and Pa B when I got home why all the kids looked white at school and they said, 'Cos they are white,' and that dark kids don't go to schools like this unless white people pay for it. They said that I'm really lucky. But when I asked them if I was lucky cos I got spat on at school and called bad names, they said I was cheeky and that I should be grateful for the nice house and food and to have parents that love me. I said my real parents loved me too and ran to my room and cried, and that's why I'm writin' to you now, cos I have no-one else to talk to. If the Burkes really loved me they wouldn't make me go back to the school tomorrow.

Thursday, 11th March

I hate it here and I am so confused. I don't really know what I'm supposed to be doin'. I have these people, the Burkes, who say they are my family, but they send me to school where the other kids just pick on me all the time. My real parents wouldn't do that. They'd make sure no-one picks on me.

And I already have two families, my real family and the Home, and now I have lost them. I have to make the Burkes my new family. I don't know if that's what it's like for everyone, but I think maybe I am the only one in my school with three families. Maybe I am lucky to have three families cos I'll have more people to love me, but I don't feel very loved right now and I haven't been hugged for ages and ages. I wish I could go back to Matron Rose or to my real mum and my real family, cos I felt like a family there.

Friday, 12th March

I was so glad the school week was over so I didn't have to see all those horrible kids again. The only time they talk to me is when they call me names. They were really nasty and I tried to tell Ma and Pa B but they just said I had to learn to live with it cos I was a bit different and that I'd get used to being there. I don't know about that. How do you get used to people calling you names all the time?

Saturday, 13th March

It's really early on Saturday morning and I can't sleep and it doesn't help that Sophie is snorin' so loud. I wish she could hear herself. So I'm sittin' at the windowsill looking out to the garden and geez it's bigger than I thought. It's like a little farm or somethin'.

The Burkes grow their own fruit and vegies and even have a couple of hens. Ma B says they have to have these things cos there's no train from St Ives to town to get everythin' she needs and Mr Johnson doesn't have everythin' in his shop. She buys the really big things she needs for the house like sheets and stuff at the Anthony Hordern's annual sale in Pitt Street, or sometimes she orders things from a magazine called a catalogue. Then they post you the stuff. I like feedin' the hens though and collectin' the eggs.

Sunday, 14th March

You won't believe what I learnt in church today. There's differences between the Catholics and the Baptists. You see, there was a little baby being baptised in church today and that never happened in the Baptist church I went to in Bomaderry, cos you get baptised when you're grown-up in the Baptist church.

When I told Sophie I hadn't been baptised she told Ma and Pa B. I think Pa nearly had a heart attack or somethin'. He said, 'That'd be right, bloomin' Protestants, can't even raise their kids in the church properly. We'll get her baptised soon. I'll talk to Father Frank about it next week.' But Pa didn't even ask me if I wanted to be baptised. I wasn't even sure that I knew what it meant. Was I gunna be a Catholic now, or a Baptist Catholic or what? I was too scared to ask though cos Pa seemed really annoyed. His glasses steamed up and he started sayin' other stuff to Ma about how all the blacks in America were Baptists and it must be a coloured thing. There was that word again, 'coloured'. What did they mean by that? We're all a colour aren't we? Like Sophie's a bit pink, and I'm a bit brown, and Pa, well he was really red in the face when he was talkin' about this stuff. That's what makes us all coloured, I reckon.

Wednesday, 17th March

Somethin' weird happened today and I don't feel very good about it. Sophie and I was walkin' home from school. All of a sudden she grabbed my hand and said we had to cross the street. She started draggin' me and I didn't know what was going on. Then she said we weren't allowed to go near 'those kind of people'. She meant the dark lady walking towards us. So we crossed the road so we didn't walk near her. But this lady had a really pretty flower dress on and she didn't look bad at all. And she had a really nice face and big smile, like my real mum. She made me think of my real mum too, though I can't remember her face that clearly, just her big warm arms and her smile. I wonder where that lady lives and what she was doin' in the Hives. She's the only other not-white person I've seen round here.

Oh, it's St Patrick's Day too, which is really big for the Irish, so we had a special dinner tonight and Ma B made the mash potato go green.

Thursday, 18th March

School was OK today cos I think I might have made a friend. Her name's Antonia and she's Italian. No-one talks to her either, cos she's a 'dago'. That means someone not from Australia, though she doesn't use that word. She says they're migrants. Johnny Jones called her 'spaghetti head' the other day. She was sittin' by herself today just like me and so we sat together but we didn't talk. I hope she sits next to me tomorrow.

Friday, 19th March

Antonia and I talked on the playground today. She said I could call her Toni for short. She lives on the North

Shore too. That's what they call all the suburbs on this side of the Harbour Bridge. Her parents bring her to school in their car every day. I think they must be rich cos it looks like an expensive car and brand new. Toni's father has an orchard like other Italian people in the area. They grow lots of fruit and sell it to the markets for people to buy. She offered me a piece of her orange, which was really nice. She said she wanted to be my friend, so we sat together all lunchtime and in class. She said I was the first Aborigine she'd ever met. I said I wasn't an Aborigine cos I felt ashamed. She said I must be a part-Aborigine cos I looked like one. That made me feel better cos she made out it was all right to be Aboriginal. I'm not goin' to start saying 'I'm a part-Aborigine' though, cos most people seem to think it's a bad thing here in the Hives.

Saturday, 20th March

I'm sittin' out the back of the house on the grass right now writin' to you cos Ma B said it was all right as long as I and Sophie cleaned our room first. We have to dust the windowsill and stuff and take the sheets off our bed for Ma B to wash. It made me remember when I lived at the Home and even when I lived with my other family in the country. I remember that we lived in a red hut with two rooms. We didn't have a chimney like the Burkes, or a proper bath. We just had a big tub to wash ourselves and all the dishes and everything in. We didn't even have hot water. We had to heat water up on the fire and pour it in the tub. Everyone else had to do the same thing. There was about 100 people living where we lived, but more people lived outside the fence where our huts were. But the manager wouldn't let them in. I thought that was really strange. Mum and Dad wouldn't let us leave either. They said that we had to get permission from the manager to leave. I didn't know what that was though cos I was just a kid.

Sunday, 21st March

Today is Palm Sunday and we got palms at church. They're supposed to remind us of when Jesus rode into Jerusalem on a donkey and all the people along the side of the road tore palms off the trees cos they was so excited. We got our palms blessed at church and I've got mine above my bed on the wall to keep me holy. Sam was using his as a fly swatter and even hit me and Sophie with it a couple of times. Ma B pulled his ear but he got free and ran away. He definitely won't go to heaven. I'm glad cos I don't want to live on Earth with him and then up there with him too.

I saw Toni at church and we talked a little bit afterwards. It's nice to have a friend here.

Monday, 22nd March

Today we did a history class in school and Miss Foster talked about when the first white people came to Australia. She said that Captain Cook discovered Australia in 1770 but that there was about 50 000 Aborigines living in New South Wales then. She reckons there's only about 10 000 now cos Aborigines are a dying race. I wonder if that means Aborigines just keep dying or are racing to die. And does that mean that I'm dying, cos I think I'm an Aborigine even though people make me feel ashamed to be one?

I didn't want to ask Miss Foster what a dying race meant but I did ask her how someone could discover a country that already had people in it. Cos the way I see it, if someone was already here when Captain Cook arrived, then those other people must have discovered it first, and Captain Cook must have been a visitor or somethin'. But Miss Foster said that Captain Cook discovered Australia, and all the books say that, so I should just listen properly and not be so bad-mannered.

Tuesday, 23rd March

On the playground today Johnny Jones chased me and when I stopped cos I was so tired he said that I should keep my mouth shut in class and that I should be glad that Captain Cook came to Australia cos he gave Aborigines proper food and clothes and houses. He said we wouldn't even have schools if Captain Cook didn't come here. I said that I hate school so that's another reason to hate Captain Cook even more cos he brought schools with him.

Wednesday, 24th March

I dodged Johnny Jones today which was good. When I got home Sophie let me play with her cat, Trixie-bell. I think that's a stupid name for a cat but that's what it is. I'd call my own cat something strong like Tiger, if I had one.

Thursday, 25th March

I met Ma B at the shop near school today. It has everythin' there: the Post Office, the newsagency and groceries like sugar, tea and flour. Ma B was gettin' some flour to make somethin' special for Easter, I think. She introduced me to Mr Johnson who owns the shop. He looked at me funny when she said, 'Mary is my daughter'. I reckon he was wonderin' what was goin' on, eh, me bein' brown and all. He gave me a lolly which was nice.

Friday, 26th March—Good Friday

We don't have to go to school today cos it's Good Friday. That's the day that Jesus died on the cross from being nailed to it. That still makes my stomach go all funny. Who would do somethin' like that anyway?

We can't eat meat today and we have to have fish.

Saturday, 27th March—Easter Saturday

Sophie was gettin' all excited about Easter today, helpin' Ma B paint eggs and stuff. I just sat around and thought about all the kids at Bomaderry. I wonder what they are doin' and what Marj is doin' in the home at Coota. I wonder if she has to still go to church all the time like I do. I wonder if she got my letter or not.

Sunday, 28th March—Easter Sunday

It's Easter Sunday and the church was all full of white lillies, white narcissuses and evergreen yew. Father Frank said it was a day to celebrate cos Jesus had risen and was alive again. Me and Sophie reckoned it was a day to celebrate cos we got some Easter eggs. She's nicer to me now. She says I sing real good too. Toni and her family was at church too, and they was all dressed really fancy. I think Italians must be really Catholic and really rich too.

Monday, 29th March—Easter Monday

It's very different here to Bomaderry. I notice it all the time. Like today I didn't have to have porridge cos Ma B said if I wanted to try somethin' else that would be all right. So I had somethin' called Cornflakes. Sophie likes Cornflakes best. It's a cereal you have with milk and sugar and it's nothing like porridge. She said I could have Weeties another time if I liked. Wow, we never had a choice at Bomaderry.

Tuesday, 30th March

I hate my school and I hate Johnny Jones. He called me names again today. He said I was dirty and that 'all abos are dirty'. Toni told him to shut up but he called her a 'filthy dago'. So I came home and had a really long tub and tried to scrub the brown off my skin in case it was dirt. All that happened was I scratched my skin and now it's red in places. Why is my skin a different colour and why do they all make fun of me? I hate Johnny Jones and his friends. I hope the devil gets him. Betta still I hope someone nails him to somethin'.

Wednesday, 31st March

I woke up this morning and looked at my skin and it's still brown, but all the red patches have gone. At least I can tell Johnny Jones that I'm not dirty, but how can I explain that I've still got brown skin?

Thursday, 1st April—April Fools' Day

Johnny Jones played a stupid trick on Miss Foster and said April Fool. She said he was a fool though and the rest of the class laughed even louder than they did when he tricked her. I laughed really, really loud and so did Toni.

Friday, 2nd April

Johnny Jones and his ugly bully friends picked on me again today calling me names and I don't know why. What did I do to them? He said I was an orphan and that my parents didn't want me and he made me think that maybe that was right. I wanna write a letter to my real mum and ask her but I don't know where to send it. Maybe if I write to Matron Rose she can find out where my mum is and send it to her.

Saturday, 3rd April

Today I wrote a letter to my real mum and I'm gunna send it with Matron Rose's letter when I write it. Here's what I wrote my mum. I hope it's all right. I haven't written to her for years and there's so much catching up to do.

> Dear Mum,
>
> How are you? And how is Dad and the other kids? Where are Busta, Max, Bluey, Betty, James, Janie and Margaret? Please say hello to everyone, please.
>
> I am not at Bomaderry any more but living with another family in Sydney, in a place called St Ives, but I called it the Hives, cos I've seen a few bees round here. Ma and Pa Burke look after me and I don't call them Mum and Dad cos you're my real mum, aren't ya?
>
> I go to St Ives Public School in Rosedale Road which is not far from the Burkes'

house, and I'm having a bit of a bad time but don't worry cos I'm actin' all grown-up and everythin'. Matron Rose did a good job makin' me a proper lady and all.

Can you write to me and tell me about the family cos I really miss you and everyone?

And Mum, can you tell me something? Do you still love me?

Lots of love and big hugs and kisses,

Mary xoxoxox

PS: I don't know why but they changed my name to Mary Talence so when you write to me you have to put Mary Talence on the envelope. You can call me Amy in the letter though cos we know that's my real name, eh? My address is 5 Putarri Ave, St Ives, Sydney, Australia.

Sunday, 4th April

On the way to church today Pa B was a bit angry and was talkin' all the way there to Ma B about how there should be a Catholic church in the area. See, there's a Methodist and an Anglican church but no Catholic church. That's why we have to drive to the Catholic church in Lindfield every Sunday. It's called the Holy Family of Nazareth and can fit almost three hundred people. That's a lot I reckon. That's more than our school.

It's an all-right church with a school and it's made of brick. At the front of the church is something like a stage. That's where Father Frank says mass, but it doesn't look like the altar like other churches I've seen. Maybe the school uses it for plays or somethin'. The back of the church has a statue of Mary and another statue of Mary, Joseph and Jesus. That's a pretty small family, I reckon, not like a real family, eh? Or even the Burkes. There's three windows at the back of the church and they're really small but there's windows all along the side of the building and there's loads of sunlight. I like that. It's really bright inside. And there's all these beams holding the roof up. I don't reckon the roof will ever fall in with all those beams. There must be about ten or twelve of 'em.

Anyway, Pa B was goin' on about havin' to drive there, but Ma B said to be positive and see it as a day out of the house. Lots of the kids from my school go there too and the music teacher, Mrs Elliott, sings in the choir. She seems all right. She says I sing real good too.

Monday, 5th April

I'm trying to be a sister to Sophie cos I'm used to havin' lots of other kids around but today she made fun of me cleanin' things. That's what we used to do in the Home. Even when I lived with my real mum I remember her cleanin' all the time cos our house used to always get inspected by the manager's wife to make sure things was tidy and clean and there was no clothes or food lyin' around. Mum used to even sweep the dirt ground outside the house to make it look clean and tidy. So I always clean the room and help Ma B with the housework cos that's what I'm used to doin'. Sophie reckons I can be her slave now, but I said I only help Ma B and she can clean her own mess. She said that I can't make Ma B love me just by cleanin' up and that Ma B loves her more cos she's her real daughter and I'm not. That's probably true but Ma B is pretty good to me anyway.

Thursday, 8th April

I've been real busy with schoolwork and housework and haven't written for a couple of days. Sorry. I'm back now.

Today Sophie made fun of me when I was all surprised at how easy it was for Ma B to do the washin'.

I told her that we had to get the men to carry all the washing down to the river when I was a little girl and make a fire to boil the water and wash the clothes over. She said my real family must be real primitive. That's a word Miss Foster uses in class when she talks about the Aborigines not being smart enough to make housework and stuff like that easier. I said Aborigines could start fires without matches though, and she said it took a long time the other way. Sometimes I don't think Miss Foster is that smart for a teacher, but I don't say that in class cos I'd get in big trouble, eh?

Saturday, 10th April

Today Ma B taught Sophie and me how to make Anzac biscuits. She said families in Australia used to send these to the Australian soldiers that went to war and that they can keep a long time cos they don't contain any milk or eggs. I'm gunna put the recipe here so I can use it maybe one day later, or maybe I'll send it to Matron Rose so she can make them at the Home for the kids there. You need:

4 tablespoons butter
1 teaspoon baking soda

2 tablespoons golden syrup
1 cup flour
1 cup rolled oats
2 tablespoons sugar

You mix the flour, rolled oats and sugar in a bowl, then melt the butter in a saucepan on the stove and stir in the golden syrup and baking soda. Add the butter mixture to the flour mixture and mix well. You then put teaspoonfuls of the mixture onto a baking tray and put them in the oven for ten to fifteen minutes. I don't know how hot the oven should be but you can tell when they're cooked and they're really yummo.

Making the biscuits with Ma B reminded me of when I was really little and we ate Johnny Cakes. They're made with flour, water, and salt and are cooked in drippin' or lard until golden brown. Then you smother them with treacle. Sometimes we just chucked the dough on the coals to cook. Yummo!

I might have to change the name of them now cos of Johnny Jones. He'll give those Johnny Cakes a really bad name, eh?

Sunday, 11th April

Today at church Father Frank talked about how Jesus died on the cross for us. I was confused cos they say Aborigines are bad and white people are good, but white people nailed Jesus to that cross so that should make them bad too. Now I'm confused about who is good and who is bad. I thought putting nails in someone would make you bad, eh? I just have to pray, cos I don't wanna be a sinner or an evil person or someone might nail me to something and that'd really hurt. I stepped on a nail once and I cried all day, and it got a bit infected.

Toni sat next to me in church today too. We saw Johnny Jones there and reckon he was probably praying for a new brain. He came up to me after mass and said that I shouldn't be praying to a white God cos I'm not white. I said I was too and anyway how did he know God was white when you can't even see him? Then Ma B came and got me cos we had to go home.

Monday, 12th April

I've been livin' with the Burkes over one month now and still things ain't right for me. I feel so lonely all the time even though I share a room with Sophie and live in this nice big house. Sam still doesn't accept me and never talks to me and neither do the kids at school except Toni, but she has some family at the school and sometimes plays with them. Ma and Pa B just say it will take some time to get used to everythin' but how much time will it take? I'm so sad and feel really left out all the time.

Tuesday, 13th April

While I was sittin' by myself at school today I decided to write a song for my mum, not just to pass the time but cos I saw some mothers at the school and it made me think of Mum. I'm still working on it but it goes somethin' like this for now:

Mum, Mum with the big brown eyes,
Mum, Mum with tears in her eyes,
Mum, Mum, with the hugs and the kisses,
Mum, Mum, I'm the one she misses.

I love my mum, and dad too,
I miss my mum's hot beef stew.
I love my mum, and she loves me
I miss my mum, can't you see?

Wednesday, 14th April

I got sad at school today cos one of the kids, Julianne, brought in photos of all her family. Her sister got married a while back and they had photos of the weddin' and all the people there. She has a really big family and they all looked really happy. I wished I had photos of my family, and I wonder if my sisters got married too. How will I know when their weddin' is and how will I get there? I hope I haven't missed out cos Julianne's photos look like weddin's are fun and everyone gets dressed up. I wonder if the Burkes have a camera. I think only rich people have them. We didn't have one at Bomaderry, but sometimes men in suits would take photos of all us kids together.

Saturday, 17th April

Pa B really loves sports and he's always talkin' about different kinds. Ma just smiles and listens cos she's a real lady and does women's stuff around the house like cookin' and cleanin' and lookin' after us kids. Anyway, Pa B was talkin' about some man called Harold Nossiter and his two sons who were the first Australians to sail a yacht, which is a boat without a motor or paddles, round the world. They sailed and got back into Sydney yesterday, but left Australia on July 14th 1935. That's a lot of sailin', eh? Pa had a beer and said cheers to no-one, but I think he was celebrating for that man and his sons.

Sunday, 18th April

We went to church today and then went to see the Sydney Harbour Bridge. It was designed by a man called John Bradfield and it's only five years old. I said to Ma B that it must be the biggest bridge in the world and she said I was right. It's almost 1500 feet long and the widest and the heaviest in the world. She also said that sixteen men died building the bridge. One other man fell into the water and lived. Maybe they fell from right up the top which is really high, about 426 feet above water level.

Pa B reckons that before there was a bridge across the harbour there used to be punts that carried cars across and went from Milsons Point to Bennelong Point and from McMahons Point to Dawes Point. Pa reckons that hundreds of houses had to be torn down to make room for the bridge, too.

We walked right across the bridge and back again and then we had lunch sittin' underneath the bridge at Milsons Point lookin' at the harbour. Pa B told us all about the bridge and when it was opened. I thought it was really amazing how they made that thing stay up there in the sky and when I said that out loud, Sam said I was really stupid, just like all 'boongs'. Pa B clipped him on the ear and said not to say that again. Sophie and I looked at each other and smiled cos Sam is always sayin' horrible things to both of us, but we can't hit him cos he hits us back and he punches much harder than we can.

Monday, 19th April

I finally wrote that letter to Matron Rose today to tell her about my new life here in the Hives. Here's a copy of what I wrote so I know exactly what I said in case she doesn't get it and I have to send another one.

> Dear Matron Rose,
>
> Here I am in St Ives. This place used to be called Rosedale but I think they changed the name cos this place looks like a place in Cornwall, England, called St Ives. That's what Miss Foster, my new teacher, said anyway. But Ma B (that's what I call my new mum here)

she reckons that it's named after a man called Ives who was in parliament and helped to get the first post office here in about 1885. Who knows? Who cares anyway?

So, I am okay, as you can see, but I really don't like it here that much and I'm always thinking of you and all the kids there. How is little John? Does he miss me? Do you miss me? And why did you let the Protection Board people send me here? I really need to ask you now, what was I doing in the Home in the first place? I don't even really know why.

Anyway, I have a brother and a sister here, but I don't think either of them like me. Sam—I don't think he likes anyone though—and Sophie who is really spoilt and isn't used to sharing things like I am, though she's all right sometimes.

Ma and Pa B seem really nice, but they keep telling me to forget my real parents and be grateful for everything I have here. They say I'm better off here with them, but better off than what or where? I don't understand.

I'm gonna have my name changed again. This time to Mary Burke, but I wanna stay Mary Talence cos that's who I am, isn't it? I don't like changing my name so much. I won't know who I am if I do it too much! And they want to baptise me. I don't know about that. What do you think?

Well, I betta go now cos I always help Ma B get ready for dinner. Sophie doesn't help her and she hates it when I do cos Ma B really likes it. I'm not trying to cause trouble. I'm just so used to helping with dinner and the dishes and stuff.

Please write back to me soon. My address is 5 Putarri Ave, St Ives, Sydney.

Love and hugs,

Mary Talence

PS: I've also written a letter to my real mum. Can you post it to her for me, please?

Tuesday, 20th April

Today we had to write a story with the title 'Who Am I?' Miss Foster said it's somethin' you call an autobiography. That's when you write about your own life and it's all truth. Not like the stories that Sophie reads like *The Magic Pudding* and *Snugglepot and Cuddlepie*. They're real kids' books anyway.

The teacher said we had to write a one-page story about who we are so we can then read it to the class so they can learn more about us. We have to write about where we come from and what we like and stuff like that, so we have to think a lot of what makes us different to each other, and even some of the things that make us the same.

This made me scared cos I'm really different to everyone else in my class. I'm always thinking, now that I live with the Burkes, who am I really and what am I doin' with this family in this big house? I don't look like anyone round here either, especially not Sophie who's really pink with freckles and straight blonde hair. I don't know where my real mum is and why she let me go to the Home. I feel like I'm livin' the wrong life all the time cos I have to pretend the Burkes are my family but they're not really. They are Sophie and Sam's family and I feel like I'm in the way all the time.

Miss Foster says we can have till Friday to finish our story and then we have to read it to the class.

Wednesday, 21st April

I worked on my story called 'Who Am I?' today and it was real hard to get started. So I started like with my diary and said my name and stuff. It has to be a whole page though so I have to write a lot more. I'm gunna go now and try and think of what else to put in it and I'll write to you tomorrow.

Thursday, 22nd April

I think I finished my story about myself and here it is so you know a bit more about me:

'Who Am I?' by Mary Talence

> My name is Mary Talence and I am ten years and three months old. I used to be called Amy Charles but they changed my name when I went to live at Bomaderry Children's Home. I have three families: my real family, my family with Matron Rose at the Home and my new family, the Burkes. My new brother Sam and sister Sophie go to the same school as me. I have long shoulder-length dark brown hair that is a bit curly,

and I have brown eyes. I am not too tall for a girl and I have skinny arms and legs and really skinny ankles like my real mum and all the other kids in the Home. My favourite colour is blue, even though blue is a boy's colour. I really like light blue, like the colour of the sky on a sunny day. I have three blue ribbons, all different kinds of blues though. I like to collect ribbons. I like to sing very much and I write my own songs and play the guitar a little. I want to be a famous singer when I grow up. I like going to church cos I get to sing out loud with everyone and I also like to write in my diary about everythin' I do every day. I like sittin' in the big park near the Burkes' house and lookin' at the birds and all the people havin' fun and the kids playin' together. I don't like brussel sprouts or carrots but I love lollies and Ma B's chocolate cake with a glass of really cold milk. I have one friend called Toni, but my bestest friend Marj is in Cootamundra Girls' Home.

Friday, 23rd April

I was so scared today that I'd have to read my story out and everyone would laugh at me cos I'm so different to everyone else. I sat there and hoped Miss Foster wouldn't pick me first and she didn't, thank goodness. She chose a girl called Susannah that I knew nothin' about and I learnt all about her and her family and her father who is a famous musician and travels around the world. She had lots of things to say about music and other countries.

Then Miss Foster got Johnny Jones up and his was really stupid. He said he liked throwin' stones on people's roofs and hittin' his sister and usin' a shanghai to hit birds. I didn't really listen to his story cos I was busy writin' a note to myself about how much I hate him.

I really liked Toni's story cos she talked about nice things like being around her family and stuff like that. She said she really liked all the holidays like Christmas and Easter and especially Valentine's Day, which I'd never heard of before. We didn't have a Valentine's Day at Bomaderry or with my real family. But she said it was somethin' from Rome—that's in Italy where her family are from. She says it's a special day for showing people you love them. It's on February 14th and Toni said it was a day to honour Juno who was the Queen of the Roman gods and goddesses. She said people give flowers and chocolates and it's really romantic, but that you can even just tell your parents you love them on that day. She says Valentine's Day is really big in Italy but not in Australia, which I'm glad about cos I can't tell my real parents I love them cos I don't know where they are, so I'd be really sad on Valentine's Day.

Then it was my turn and I was really nervous, but no-one said nothin' about my story. Maybe they was all tired or bored from listening to the others. Who knows? Johnny Jones said I was greedy havin' three families but I told him I didn't have any say in it. I just did what I was told to do and went where I was sent.

Sunday, 25th April—Anzac Day

Today's Anzac Day and there was a march in town for all the men who went to fight for Australia in the war. ANZAC means Australian and New Zealand Army Corps and it's the name that they gave the soldiers who fought at Gallipoli in Turkey. Pa went to have some beers with some of the old men from round here who were in the war. When he came home I think he was a bit drunk and he made us all stand in the kitchen and say this thing he read off a piece of paper. It said:

They shall not grow old, as we that are left
    grow old,
Age shall not weary them or the years
    condemn.
At the going down of the sun, and in the
    morning,
We will remember them.

Then we all had to say 'We will remember them'. Pa went to bed then.

Monday, 26th April

Miss Foster was talkin' to Ma B today at school and they looked really serious and sure enough when I got home Ma and Pa B sat me down and told me what Miss Foster said. She told Ma about my story 'Who Am I?' and that I talked about the Home and my real family. Ma and Pa weren't really angry but they said I had to forget about the past and live in the present and that they didn't want me to mention Bomaderry or my real parents ever again. They said it made me look ungrateful. 'Are you that ungrateful that you don't appreciate what we do for you, Mary?' That's what Ma said and she was nearly crying. I felt really bad and promised not to say anythin' any more about Bomaderry. Ma gave me a hug and Pa rubbed the top of my head, and said I was a good kid even for a fair-skinned Aborigine. Ma told him not to say that again, that I was learnin' to be a respectable white person, and that was good. Now I know why they sent me here, so I could be white and not be Aboriginal any more cos being Aboriginal is something to be ashamed of. For the first time ever I really wished I was white like everyone else around me.

Saturday, 1st May

We had a really good day today cos Ma B took me and Sophie and Sam on the train from Gordon to the city and we went to the children's library at the Sydney Municipal Library. It's been open for twenty-five years already and there was lots of books. Sophie really liked lookin' at *Mary Poppins* though and sat there readin' it almost the whole time.

I walked around all the sections and looked at some books about Sydney and stuff and read about where we live in St Ives. With the other suburbs like Pymble and Killara, Lindfield, Roseville and some others, the area's called Ku-ring-gai. It's named after the Aboriginal tribe that lived there before white people landed. The Ku-ring-gai people owned the whole area of North Sydney, I think.

And you know what, St Ives stands on the edge of Ku-ring-gai Chase Park and everyone's been calling it 'The Chase' since 1896. Then in 1928 the area was called Ku-ring-gai Municipality. There's supposed to be Aboriginal rock engravings in the park too, wonder if I'll ever get to see 'em. I reckon I'm gunna call the area 'The Chase' from now on.

Sunday, 2nd May

We had the best day today. I didn't want to come home. We all went by ferry to Taronga Zoological Gardens which is right on the harbour. Pa reckons it was opened in 1916 and they're gettin' new animals all the time.

I don't know what I liked the best. I really liked the elephant ride with the man in the cap drivin' the elephant. There was about eight of us on it, and we nearly got stuck goin' through an archway. But I also liked the children's kindergarten with the rabbits and I almost fell over the fence tryin' to feed them. Lucky I didn't, eh? We all went on the train ride too but not Sam cos he said it was for little kids. We saw the big giraffe and when he looked up you could see the Sydney Harbour Bridge behind him. It looked really small cos it was a long way away. We went to get some refreshments at the Tea Rooms before goin' to see the lions, tigers and reptiles. The tiger didn't move or nothin' but the lion roared really loud and it scared Sophie but I just laughed. Before we left we checked the time on the floral clock, which is a round garden of flowers with hands like a clock. Really weird. I wonder how they do that, eh?

It was a great day and the only thing that ruined it for me was Sam sayin' that Aborigines were like orang-outangs. There was three orang-outangs at the zoo called Freda, Freddie and Peter. Sam goes, 'Hey Mary, there's your real family over there,' and pointed to them. I just wanted to cry and I think Ma and Pa B saw my face. While Ma put her arm around me, Pa clipped Sam across the ear. That almost made the insult worthwhile. I don't know why he says such terrible things all the time, and where does he get ideas like that anyway? Probably cos no-one ever says nice things about Aborigines. That's cos they don't really know any, I suppose. I was born Aboriginal but they want me to forget about it and pretend to be white all the time.

I'm gunna try and not worry about it any more tonight and just remember the nice things about today like the walruses and seals jumping for fish and splashin' in the water.

Monday, 3rd May

It's nearly three weeks since I wrote to Matron Rose and she hasn't written back. I hope she's not angry with me for asking those questions. I hope she got the letter.

Tuesday, 4th May

Still no mail from Matron. I never heard from Marj neither.

Wednesday, 5th May

Still nothin'.

Thursday, 6th May

I'm so excited that it arrived today, my letter from Matron Rose. She sent a photo of the kids too. Someone musta been there takin' photos. Anyway, I ran home from school and checked the letterbox and there it was. It said 'Miss Mary Talence' on the front and I opened it straight away. She says:

Dear Mary,

It was so lovely to hear from you, and your writing is getting much better. I'm glad the Burkes are taking care of you. You will settle into school very soon. Don't listen to those bad boys and girls. They won't go to heaven if they behave like that all the time.

All the children miss you very much, including little John, even though he's still a baby. I'm sure he knows that you're not around, but he's being a very good boy, so you must have done a good job training him to behave properly.

I didn't want to send you away. You are the best girl I have had here at the home, but the Aborigines Protection Board said you had to leave. Please don't be angry with me. It is much better that you live with the Burkes because they can help you have a good life like white people and have all the things white people have.

Write me a letter again soon and tell me all about your school work and the friends you make.

Best wishes,
Matron Rose (Mum)

She didn't say anythin' about my mum's letter. I hope she posts it to her. I'll have to write her another letter to make sure. I'll do that another time. I'm just so excited about hearin' from her.

I didn't even tell Ma and Pa B about it cos I know they'd get upset or somethin'. They might think that I don't like them and that's not true.

Friday, 7th May

You'll never guess what? I met that woman in the shop today, the one that Sophie and I saw weeks ago and crossed the road to get away from. Her name's Dorothy, or Dot for short. She started talkin' to me straight away cos she said I was Aboriginal. But I said I was white and she looked at me with a weird frown on her face. It was really strange cos she's the first dark person I've seen here since I arrived. She said that we are the only ones in the whole suburb, and that she was glad to meet me. She told me about the house where she works. She's a domestic servant for a family called the Sullivans and she's been there for three years. She reckons that all the people that live in her street are doctors or dentists or lawyers. Her house is a big red brick one with leadlight windows and a really fancy garden with roses and nice grass.

I couldn't talk long cos I had to get home. Ma B always says don't dawdle home from school and don't talk to strangers, but Dot talked to me first and seemed really nice so I think that might be all right. Dot said we would see each other again soon. When we got to my street she said that the word 'putarri' was an Aboriginal word for 'cold'. That didn't surprise me cos I always feel cold here, like there's not enough love or somethin'.

Anyway, the bad thing is that when I came home and told Ma B about Dot she got really angry and said I wasn't to talk to her again, not cos she was a stranger but cos she was an Aborigine. She said I wasn't allowed to mix with those kind of people. I don't know what she meant cos Dot said I was her kind of people. I'm really confused and sad cos it was the first person who talked to me like a friend since I got here. I hope Ma B changes her mind. I'll ask her again in the mornin'.

Saturday, 8th May

Ma B was really quiet with me today so I didn't say anythin' about Dot. I think I should just leave it for a while.

Sunday, 9th May

Today we took a drive in the Burkes' car. They bought it on higher purchase, and I thought it meant buyin' somethin' at a higher price instead of a lower price, but Ma B said it's when you pay for it a bit at a time. You can take the car home when you make the first payment. It's really called hire purchase.

Anyway, the car is a Lincoln, not like the American Lincolns, as Pa pointed out to me. He said it has a European engine with American parts so it's an international car. It's already ten years old though so I reckon by the time I can drive it will be too old, eh. But I like it cos it's one of the cars where the roof comes down and you can have the wind in your face and hair. Sophie hates it cos she reckons it messes her hair up. She's so prissy sometimes. Anyway, it was rainin' today so the roof was on properly and we just had to look out the sides at the scenery. We drove to Manly Beach and watched people bodysurfing which was really excitin'. There was really big waves from the storms but the men in the water made it look real easy. I don't reckon I could do it. Sam said he could, of course. I'd like to see him try, and get all churned up by the waves!

We got out of the car when the rain settled and we

walked along the promenade which is lined with these big pine trees like in a forest. It looks really nice and green. We had an ice-cream too and Sophie and I had a lickin' competition. I don't know who won though.

Monday, 10th May

Well I asked Ma B about talkin' to Dot and she got really angry with me again and got Pa B. He said the same thing. I was not allowed to talk to any Aborigines, and I was supposed to try and be more white, like Sophie and Sam, cos good people were white and bad people were the others. They didn't tell me what they meant by the good and bad people though. I don't want to be a bad person, but I look in the mirror and I don't really look white. All my real brothers and sisters are brown too, and so are my real parents, so how can I be white? And what about Dot? Is she trying to be white too? She can't though, cos she's real dark. If I see here again I'm gunna ask her.

Tuesday, 11th May

After school today Ma and Pa B reminded me again that I shouldn't talk to Aboriginal people cos they are dirty people who drink too much alcohol. But I tried to tell them that Dot was really clean, with white gloves and everythin', and that I reckon she probably goes to church and prays to God too, like a white person. I'm sure she doesn't get drunk. They just said that Aborigines are not good people and that I was to do as I was told or I'd end up like the rest of them—Aborigines I think they meant.

Wednesday, 12th May

There was a special announcement at school today about the coronation of King George VI, that means the sixth. A coronation means that he was crowned. Matron used to say that if we played up she'd crown us. I think she meant something different though. I miss Matron Rose a lot.

Thursday, 13th May

Mrs Elliott gave us a project in music class today. We have to put together something musical and perform it to the class. You can write a song and sing it or play an instrument or somethin'. Toni plays the violin which is really hard to play but she's going to do somethin' on that. I'm gunna write a song, but I've gotta think about what I'm gunna write about. Maybe a song about me. I need to think about it and will get back to ya.

Friday, 14th May

I've been thinkin' about my music project and I think I'll write a song. So far I have these words and I'm using Sophie's tambourine, only cos Ma B said it was all right. Sophie didn't really want to lend it to me.

I live in St Ives, I live in St Ives,
There's lots of bees, but not many hives,
The ground is flat and there's lots of trees
I live in St Ives.

I live in the Chase, I live in the Chase
It's a pretty nice place.
People are white and they seem all right
I live in the Chase.

Saturday, 15th May

It was rainin' really hard all day today, just pourin' down. Pa B said it was rainin' cats and dogs, but I didn't see none. So we all just stayed inside readin' and stuff like that. I practised my song and I had to let Sophie play the tambourine cos it's hers, and I had to change the words a bit. You know why.

I live in the Chase, I live in the Chase
It's a pretty nice place.
Everythin's bright, except at night,
I live in the Chase.

Sophie seemed to like the song.

Sunday, May 16th—Ascension

After church today we went for a picnic at Bungaroo, which is the big park in St Ives. It's like a forest. It was a really sunny day, with that blue sky that I like. I even had my sky blue ribbon in my hair so I matched the sky. Pa B was talkin' the whole time we was walkin' through the trees and when we stopped he said that Governor Phillip stayed there for a night on April 16, 1788, just after arrivin' in Sydney. Pa reckons it's a really significant site cos of its history with the First Fleet people. I asked what happened to the Aborigines that were there before Captain Phillip arrived. You know, those Ku-ring-gai people I read about. Pa didn't say anythin'. He just sucked his tongue to the top of his mouth and made that funny sound that parents do when they're not happy with you. The picnic was nice anyway.

Tuesday, 18th May

Wow, today my letter arrived from Marj. Matron must have posted it up to me. I ran home from school and there it was it the letterbox. Lucky I got the mail first cos maybe Ma and Pa B wouldn't have let me have it seein' as I've gotta forget about the Home and all that. Anyway, here's what she said:

To my best friend Mary,

Here I am in Cootamundra Girls Home, and I hate it. It's not like Bomaderry, I don't have a best friend or no-one like that yet. I really miss you and all the little ones there now.

It's really cold and miserable here, nowhere near as nice as Bomaderry. But it's right on top of a hill so you can see all around Cootamundra from the garden.

The building is different here too, because it's not cottages like at Bomaderry but it's an old hospital. You can imagine how warm that feels. We have dormitories to sleep in and they're the old wards. The other night this girl Ellie, who's only ten, kept screaming when we were all trying to sleep. The staff, who are all white, came and took her away. No-one really knows where she's gone, but someone said she'd gone to a mental hospital. This place probably drove her crazy. I hope she's all right.

When the girls want to go to the pictures or something, they can do some gardening to earn pocket money. I haven't been yet. I just do the normal work like scrubbing, washing, ironing and sewing. It's like being a slave, you know. And I'm not even eleven years old. Geez, they say when we're fourteen or fifteen we have to go and live with a family and do all this for them, too. Someone should be doing it for us I think, because we're only children really.

One thing that's the same here is that they tell us that our parents don't love us or care about us. But you and me know different. They must think we're all stupid. Some of the others here believe it though, because there's hardly ever any visitors, but sometimes we get mail like my letter from you.

Anyway, I better go now. I hope you are all right. Please say 'Hello' to everyone in the Home for me and tell them I miss them all.

Write soon,

Your best friend Marj xoxoxoxox

Wednesday, 19th May

I'm still thinking about what Marj wrote in her letter yesterday. Coota doesn't sound anything like Bomaderry and it makes me feel bad cos I've got this family and all these things and Marj is doing all that hard work. Could you imagine if someone like Sophie had to go and do that? She doesn't even make her own bed. There's no way she'd sew her own clothes, and especially not anyone else's clothes!

Thursday, 20th May

We had to do our music projects in class today and I did my song about St Ives and used the little guitar and everyone clapped. Toni was the only one who played the violin and she was great, but Johnny Jones tried to play the recorder and he sounded really stupid.

Friday, 21st May

After school today Ma B took me and Sophie to have our hair cut. We both just had a little bit cut off cos the hairdresser said that long curly hair was all in style. Sophie was sooky cos she's got dead straight hair, but when I told her that I could curl her hair for school she was much happier.

Ma B didn't get her hair cut. She wears hats nearly all the time anyway. Usually she has them on one side of her head and she looks really pretty cos she wears lipstick and paints her fingernails. I asked Ma if lipstick was for bad women cos of what Matron said, and she just frowned. She's always fixing her make-up in the little mirror she carries in her handbag. She even has a fur coat made from rabbit skins that she wears on special occasions, like when her and Pa B go out with friends.

Saturday, 22nd May

You won't believe this, Sophie demanded I curl her hair today, without even sayin' please or nothin'. When I said 'No' cos she was so rude, she said she'd tell Ma and Pa B that I was talkin' to Dot at the shop. I wondered when she'd say somethin'. So, we made a deal, that every time I saw Dot, I'd have to curl her hair the next day for school. At first she wanted me to curl her hair every day, and I said no-one would believe that I liked her that much, and she said, 'All right'. Geez, she's a sly girl I reckon, but we both win cos I get to see Dot, and Sophie doesn't look so bad with curls.

Sunday, 23rd May

I got baptised as a Catholic today and Ma and Pa B was both really happy. We even had a bit of a party after church but I don't really know what we was celebratin'. So, now I must be a Baptist Catholic or somethin', eh?

Monday, 24th May

I saw Dot again today at the shop. She was buyin' bread and milk and eggs for the Sullivans, and she had a really pretty dress on. It made her look really fancy. I told her I wasn't allowed to talk to her cos the Burkes said so. She asked me why and when I told her she got angry.

Dot asked me if I knew who my real parents were. That threw me cos then I really wanted to talk to her even though I wasn't allowed. She took my hand and we walked out the shop towards the big forest-like park. I let her even though I knew Ma and Pa B would be cross. She said that she had lots of things she could tell me if I wanted to learn, but that she had to get home to do the Sullivan's ironin' and make their dinner by 5 pm. Dot says she is a 'slave' to the Sullivans and that she does all their washin', ironin', cleanin', sewin' and cookin'. She says she hardly gets paid any money. She says she even has to get on the floors and scrub them every week. And when it's time to eat dinner, she has to eat in the kitchen by herself after she serves the food to the family. She says it makes her feel like a dog not to be able to sit at the table with everyone else eatin' the food she has cooked. I feel sorry for her and am glad that Ma and Pa B let me eat with them.

We planned to meet at the shop on Friday straight after school. I'll have to come up with a story so Ma B doesn't find out, but we're always told to tell the truth and that lyin' is bad. I think Ma B would be really angry if she knew I was seeing Dot. But I think she would like Dot cos she's really pretty and wears nice dresses and speaks proper too.

Dot said she'd meet me at the shop on Friday.

Tuesday, 25th May

Sure enough, I had to curl Sophie's hair today. I must say she looked really pretty when she went to school.

Ma said it was nice we was gettin' on so well. If only she knew, eh?

Ma and I wrapped Sophie's presents tonight cos it's her tenth birthday tomorrow. We're goin' to be the same age for a couple of months which is pretty weird, eh? Ma and Pa B got her a really expensive doll which I know Sophie will love. I got her something too, but you'll have to wait till tomorrow to see what it is.

Wednesday, 26th May

The house smelt real warm and chocolatey tonight cos Ma B made her best chocolate cake for Sophie's birthday. She really loved her copy of *Mary Poppins* I gave her. Ma gave me the money to buy it. She said it wasn't too childish and that she liked that it was written by an Australian woman called Pamela Travers who came from Queensland which is a long way from Sydney. I thought that maybe cos she read it at the library that time that she wouldn't want it, but she said it's her all-time favourite book and she's really happy to have her own copy. We had a special birthday dinner of roast lamb with peas, pumpkin and gravy. Yummo.

Thursday, 27th May

Sophie's still reading *Mary Poppins* to her new doll. She must know that story off by heart by now.

Friday, 28th May

I counted the minutes down to see Dot today. She was waitin' at the shop when I got there and I ran all the way from school, and was really hot by then cos the sun was just streamin' down, even though it's nearly winter. Dot seemed like she was really anxious to talk. I don't think she has many friends round here, cos there aren't a lot of Aborigines in Sydney I don't think, or you just don't see 'em near St Ives.

Anyway, she told me all about herself. She's 17, and her whole name is Dorothy Helen Murray. She says she is from Cowra and went to Cootamundra Girls Home when she was eight and then left when she was fourteen. I asked her if she knew they called it 'Coota' and she laughed and said yes.

Dot says she is an Aborigine and that her tribe is the Wiradjuri tribe from central New South Wales. I had to get her to write the name down cos it was too big a word for me to remember and I didn't know how to spell it. You say it like this: Wih-rad-jury. She asked me if I knew my tribe. I said that it's bad to be an Aborigine cos that's what everyone around here tells me. I asked her if she was ashamed to be an Aborigine. That made her frown for a long time until she said, 'You're Aboriginal, Mary, and white people want you to forget that, cos they're trying to make us like white people.' She used a really big word starting with 'A'. I can't remember what it was. And I didn't really understand what she said and she scared me a little cos she got angry.

She said there was a big meeting last month where the government said that part-Aborigines like me had to fit into the white society. That's why I was living with the Burkes. She said it was the government policy. She said that really big word again starting with 'A'. Anyway, this policy is to make Aborigines act white. I just shrugged my shoulders cos I didn't really know what she was talking about.

Anyway, she kept talking while I watched Sophie, thinkin' I'd have to curl her hair in the mornin'. That's all right though. I really don't mind that much.

She went on and I was a bit confused—well a lot confused—and she told me about when she was taken from her family. She reckons she was herded up with all the other Aboriginal children, like a mob of cows or sheep or somethin', and they were all then lined up in order of how dark their skin was, like we used to do at the Home. She said there was some called half-caste and quadroons and the real, real dark ones were called full-bloods. I wonder if that meant they was full of blood. And what about the one's that wasn't full-bloods. Were they only half full of blood? This was all new to me. I thought there was just good white people and bad Aboriginal people, but that we all had enough blood. Not this half-caste and full-blood stuff. I was too embarrassed to ask her though, cos she seemed a bit angry that I said that it was bad to be an Aborigine. I hope she talks to me the next time she sees me.

Saturday, 29th May

Didn't do much today, just cleaned our room, and lay around readin' old copies of the *Australian Women's Weekly* that Ma B has collected since 1933. I don't think she throws anythin' out cos there's a lot of them here, and some of them are dusty. Oh, that reminds me, I wonder how little Dusty is at Bomaderry. I miss him bein' cheeky.

Of course, Sophie's walkin' around with curls today too.

Sunday, 30th May

I'm still thinkin' about all that stuff Dot told me the other day and I wished I had someone to talk to about it. I prayed in church today for all those people who aren't full of blood.

Monday, 31st May

Miss Foster gave us a special project today where we all have to draw our family tree. It's supposed to have our grandparents on there and everything, but I need help with mine cos I don't know their names, or I can't remember cos I was so young when I went to Bomaderry. Miss Foster said to start researching at home but how can I do that when the Burkes aren't my real family? They don't know my real mum and dad and they won't let me talk about them either. I wonder if I see Dot again if she'll help me.

Tuesday, 1st June

Just when I stopped thinking about all that weird stuff, Dot told me about tryin' to make Aborigines be white and blood people and everythin'. She told me other stuff today and it made me really angry and confused about why white people behave like they do.

Why does the government tell white people that it's okay to be nasty to Aborigines, even if they're good people like me and Dot? She reckons that the government has something called the Aborigines Protection Act in New South Wales, which is the state we live in. Well, this Act means that some Aborigines who have 'mixed-blood', that means some white blood and some Aboriginal blood, can get these certificates to say that they are no longer Aboriginal but to get one they have to promise to stay away from their families and communities. And it means they won't be treated like an Aborigine at all cos they'll start being white. That means they'll be allowed to drink alcohol, and can sit with white people in the pictures and will be allowed in public swimming pools and stuff like that. Aborigines can't do these things, you see.

There's also somethin' called the Aborigines Protection Board that's supposed to protect Aborigines, but I don't know what they are bein' protected from. Are they protectin' Aborigines from other Aborigines? I reckon Aborigines need protectin' from all the white people makin' these laws and stoppin' Aborigines just bein' themselves. They make Aborigines feel ashamed of bein' Aboriginal and that's not right. I reckon white people and the government should be ashamed for making

Aborigines feel bad about themselves and stoppin' them having the same things as white people. Aborigines don't even have the right to vote in Australia and they was here first!

Wednesday, 2nd June

I was nearly late for school cos I had to do Sophie's hair and she was bein' really painful about it. Then Miss Foster got us to start work on our projects in class and Johnny Jones said out loud that I couldn't do it cos I was an orphan and belonged in an orphanage cos my real parents didn't want me and couldn't take care of me. Miss Foster didn't even do nothin' to him, smack him or anythin'. I started to cry cos I thought maybe he was right or somethin' and maybe that's why I was in the Home with Matron Rose.

Miss Foster told me to wash my face. Then I had to sit down and try and do the project.

I just started with me and I put 'Mary Talence' on the page next to all my brothers and sisters from my real family on one side and the Burke's on the other. I also had to put Matron Rose on there too cos she's like my mum too. But I'll see what Dot says when I see her next time.

Thursday, 3rd June

Today we all had to read out loud in class that famous poem 'The Man From Snowy River' by Banjo Paterson. What kinda name is Banjo anyway? His real name is Andrew Barton Paterson, but they call him Banjo. It'd be like callin' me 'Tambourine Talence'. Hey, that doesn't sound too bad, does it? Maybe when I'm a famous singer that's what I'll call myself, 'Tambourine Talence'.

Anyway, I thought it was a pretty boring poem about a man from Snowy River who can ride horses and round 'em up good too. There's no talk of women in the poem though. Didn't women ride horses back in them days? It's really old you see, cos it was published in 1895 Miss Foster said. You'd reckon there'd be some newer books and poems we could read, wouldn't ya?

Friday, 4th June

I met Dot in the shop again today and I told her about my family—well all I could remember, like my brothers' and sisters' names and how old they were compared to me. She put her hand over her mouth and said she thinks she knows Janie and Margaret, and that they were at Coota with her. They were close and used to talk a lot. She says they would always talk about their brothers and sisters and how much they missed them, and that Mary was their favourite. That made me happy.

She also said that my family were from Cowra too. That made me Wiradjuri. Maybe we could be related. That made me feel so happy inside to have some real family here in St Ives. I think Dot was really happy too cos she hugged me and called me 'tidda', which she says is an Aboriginal word meaning sister. But then it was getting late and I knew Ma B would be really angry if I didn't get home soon. So we left and said we'd see each other next week, cos I still need help with the project. I really liked that hug that Dot gave me today cos I don't get them any more. I hope she gives me another one next time I see her.

Saturday, 5th June

It seems like I'm doing Sophie's hair a lot, but when I do it reminds me of the time I spend with Dot and all the stuff she tells me. I'm learnin' a lot from her, more than from Miss Foster I reckon.

Monday, 7th June

I met Dot today at the shop. She gave me some lollies which was really naughty cos we aren't supposed to eat before dinner, but geez they was nice. As we walked home I asked her about helping me write my family tree. She said she would and I showed her how much I'd done.

Dot says it's important to know who your parents, grandparents and brothers and sisters are, cos that helps you know who you are yourself. Dot's really smart I reckon.

She said lots of my real family live on a mission in Cowra. When I asked what a mission was, Dot said that when white people didn't like Aborigines livin' in the towns, they rounded them up and made them live in special areas of land that weren't worth much money to the government. These places where only the Aborigines lived were sometimes called missions, which were run by the church, or reserves, which were run by the government. Sometimes people had to get special permission to leave the missions and reserves cos Aborigines had managers and didn't have that much freedom like white people in Australia.

Dot took my family tree home to think about it. We have only a few days for the project so I'm gunna see her on Wednesday and she's gunna go through it with me again. I'll write it in here for you to read too so you know who my family are, eh?

Tuesday, 8th June

Toni said to me at school today about how pretty Sophie looked with curls and I just smiled. She showed me her family tree too and I really liked it cos it has all these really different names like Luigi, Mario and Guiseppe. Her family arrived here in the Hives in 1926. Her dad was one of the first Italians to come to the area. He has a big block of land in Yarrabung Road where he grows fruit.

Wednesday, 9th June

Well I saw Dot today and it was real hard doing the family tree cos I could remember Max, Busta, Margaret, Janie, Bluey, Betty and James, but I didn't know Mum and Dad's names cos I always called them Mum and Dad, and Dad always called me his little princess. I must really be a princess cos Matron called me her 'coffee-coloured princess' too. Anyway, I got real sad cos I knew Ma and Pa Burkes' names. She's called Moyra and he's Patrick. I didn't know my grandmum's or granddad's names either, so my family tree was really not finished.

Dot says it's not my fault that I don't know all this stuff, cos if the government didn't make me and all the other Aboriginal children leave their parents, then everyone would know who their families were and where they fit into everythin', and everyone could do their family trees too.

Thursday, 10th June

I took my family tree to school today and Miss Foster said it looked all right the way I had done it, but she said that I should take off my other family, the ones that aren't the Burkes, and that I should write 'foster' family next to their name at the top. But how can I leave off my real brothers and sisters and the people at Bomaderry? I said I would cos I didn't want her telling the Burkes cos I'd get in trouble again. She's the teacher so she must be right, eh? I have to finish it tonight and take it back tomorrow. I had to ask Ma B what her parents' names were so I could put them on the family tree. This is all really new to me too. When I was at Bomaderry, no-one really talked about their families. It was like a silence about where we come from.

Here's my tree properly. Not the one I'll take to school.

Mary Talence's Family Tree

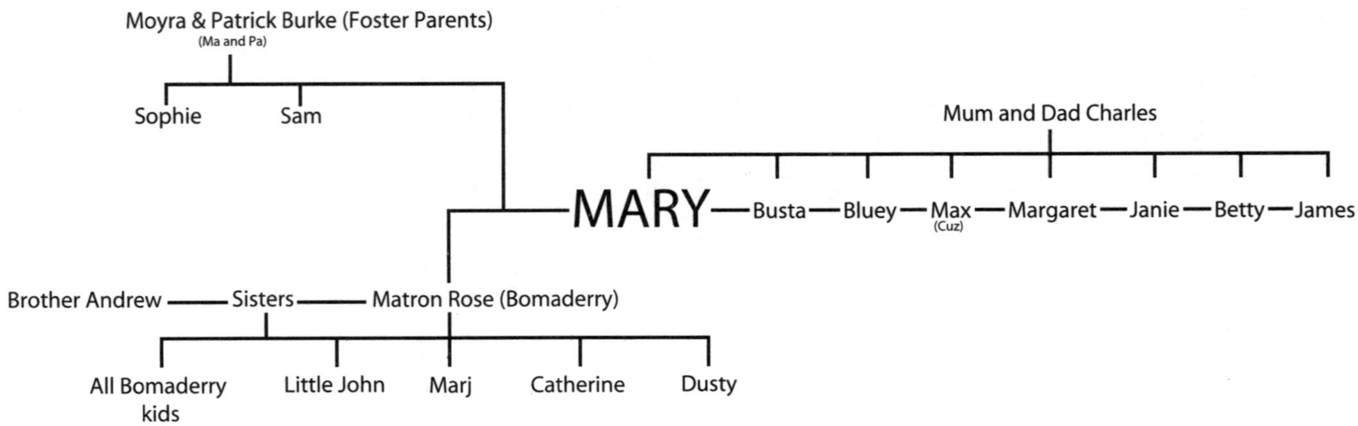

Friday, 11th June

Miss Foster told us today that we don't have to go to school on Monday. It's the Queen's Birthday and we get a holiday. We never had this at Bomaderry I don't think. She reckons it's always been a holiday on the second Monday in June, since the convicts first arrived. That means we get a long weekend, eh?

Sunday, 13th June

After mass today, Ma B went up to Mrs Elliott from school who sings in the choir at church and told her that I want to sing in the choir and the school singin' group too cos I have the best little voice. Mrs Elliott said that'd be fine and that they have rehearsals at the school on Tuesday afternoons after school with about fifteen school children. The excitin' thing is that at the end of the year they do a school play and a Christmas pageant with Christmas carols. Everyone gets dressed up like when Jesus was born. That sounds fun to me.

Monday, 14th June

It's so good that it's a holiday today. No school. Pity it was freezin' cold with frost on the ground when we woke up this mornin'. Didn't do much, just mucked around with Sophie. It was rainin' too so we stayed inside. She played with her dolls and I tried to read 'The Man From Snowy River' again, just cos I was so bored. Ma B had the book called *The Man from Snowy River and Other Verses* which is the same book Miss Foster had. Ma B loves books. I think Pa would if they was about sports.

Tuesday, 15th June

I saw Dot today and she said, 'Hello, mima,' and I thought she got me confused with someone else. She told be that 'mima' was the Wiradjuri word for star, and that she'd call me 'Mima' from now on, cos I was the star of St Ives. I told her I called it the Hives and she laughed. She's gunna teach me lots more Aboriginal words too. I can't wait. She gave me a big hug when she said goodbye. That made me feel all warm inside.

Wednesday, 16th June

I sat in the sun at lunchtime and it was so warm on my face it nearly made me go to sleep. Even with all the noise of the kids playin'. I had to keep myself busy so I didn't nod off. I wrote a song with my new word 'mima' in it. Not like 'Twinkle, twinkle, little mima' cos that sounds really stupid, but it goes like this:

I'm the mima of St Ives
I'm the apple of my daddy's eye
I'm the one with all the style
I'm the mima all the while
The other girls play with dolls and toys
I'm better than all the boys
I'm mima, the star of St Ives.

I sang it to Toni and she really liked it too. I think she might be my new best friend, but I'll never forget Marj, cos she's my bestest friend in the whole world.

Friday, 18th June

Sometimes I really hate Miss Foster cos she isn't really that smart for a teacher, you know. Today she was talkin' in class about the important moments in Australian history and she said that 1901 was important. We became a Federation and then in 1902 women in New South Wales got the vote. I said, 'No Miss, white women got the vote, but Aboriginal women didn't. Not even Aboriginal men have the vote yet, Miss.' Well she got really angry cos I made her look stupid in front of the class, but I didn't mean to. I just needed to tell her she forgot about Aborigines. Johnny Jones just laughed and said, 'Aborigines couldn't read the voting papers anyway, so who would they vote for?' That made me really mad. I said all the Aborigines I know could read and write, probably better than he can. I asked him to spell 'discrimination' then and, guess what, he just shut up cos it's too big a word for him. See, Dot was right, she told me once that just cos someone talks the loudest, doesn't make them the smartest or the rightest. (Is that a real word? Rightest? Well I'll make it a real word from now on.) Toni wrote me a note and passed it over. She reckons I'm smarter than Miss Foster.

Saturday, 19th June

Pa B came home really happy tonight cos he went to the final of the rugby league at the Sydney Cricket Ground and his favourite football team won, the Eastern Suburbs Roosters. Pa B musta won a bet too cos he brought home sweets for everyone and a coupla bottles of beer. He likes to have a beer on Saturday and Sunday afternoons. Ma B only ever has a shandy. She always puts lemonade in her beer. That's what ladies do, she says.

Sunday, 20th June

I did Sophie's hair for church today. She was really surprised cos I was just doin' it to be nice, not cos she had to keep her mouth shut. She held my hand in church which was nice too.

Tuesday, 22nd June

Today I had my first singin' group practice after school but cos it's not a Catholic school we did other songs. Mrs Elliott was really nice to me especially when she heard me sing real loud. She says I've got a good set of lungs. There was a really nice boy there but I don't know his name. I think he might be the year below me, in Sophie's class maybe.

Wednesday, 23rd June

I was askin' around at school today about that cute boy in singin' group, but cos hardly anyone talks to me it was hard. I got Toni to do some research. More people talk to 'dagos' than 'abos'. She reckons his name is Joey and he's Italian too. He's really cute with big blue eyes and really white teeth. I like people with good teeth. We all had good teeth in Bomaderry and Dot has too.

Friday, 25th June

We got our school reports today. Miss Foster wrote that I was a satisfactory type of girl and that I was a good worker and doing all right. She said that she had thought I would be subnormal cos I was a part-Aborigine, but that I was practically white in the way I worked. She even said that I was smarter than some of the other kids in class. I guess I must have made her wrong, eh, cos my work was all marked good so far. Even the music project with the song I wrote. The teacher told Ma and Pa that I like playing the guitar and that I have a lovely voice.

The only bad thing is that Miss Foster said that I don't really talk or write properly and that I don't put 'g' on the ends of my words. So, I have to try and write 'going' and 'something' and stuff like that and then I reckon I'll be a perfect student, eh? Oh yeah, Miss Foster also said I say 'eh' too much too. But no-one's perfect, eh?

Monday, 28th June

I'm sitting in the window right now with just the light from the full moon. Everyone's asleep so I can't put the light on. It's really cold tonight and it made me think about the kids in Bomaderry. I used to make sure they had enough blankets and that they was warm enough. I hope they're all right. I hope one of the bigger girls is being the big sister now. That's really important cos I think everyone needs a big sister. Now I'm here and I'm Sophie's big sister but I don't know if she really wants one though. I think sometimes she really likes having me around, and other times she doesn't like having to share things.

Tuesday, 29th June

I saw Dot today which was nice. Ma B had gone into town and I knew she wouldn't be home till late so Dot and I went to Bungaroo together and took Sophie with us. She just sat there and read *Mary Poppins* again. Sometimes I feel like I'm going to school after school cos Dot's teaching me so much stuff that the teachers don't know about Aborigines. Miss Foster just thinks Aborigines are dirty and useless. That's what it seems like anyway. But Dot reckons Aborigines are the smartest people in the world cos they invented the boomerang, and it can fly through the air and come back or it can be used to kill animals for food. Dot's real smart too. I wanna be just like her when I grow up. But she says I should get a better job than cooking and cleaning for a white family cos she hardly gets any money at all, and can't have friends over for dinner. She doesn't have much time for herself either, I don't think, except for when she sees me.

Wednesday, 30th June

I saw Dot again today at the shop and she asked me if I wanted to go to the pictures with her on Saturday. I asked Ma B and she said no cos we already had family things to do and she got real angry with me for even askin' in the first place. But I thought maybe she might have changed her mind after all this time.

After dinner I heard her talking to Pa B and she was saying that I should stay away from Dot and other Aborigines cos they will be a bad influence on me. I don't know what she meant though. Dot is a good influence on me, teachin' me stuff. And I think she is really smart.

I'm real sad and don't know how I'm gonna tell her when I see her at the shop next time. Maybe Ma B will change her mind by then. I'll say a little prayer tonight, that might help. I'll use Matron Rose's Bible she gave me when I left Bomaderry.

That's gunna be two days in a row I have to curl Sophie's hair. I hope no-one starts to ask questions. At least she's happy with the deal.

Thursday, 1st July

I had a terrible time at school today. Someone stole Miss Foster's purse from her bag under the desk. I've seen her purse. It's red with a silver clip, but I didn't take it. Miss Foster was even cryin' about it she was so upset. No-one owned up and a couple of kids whispered that I did it. Johnny Jones said real loud, 'Yeah, abos steal cos they aren't smart enough to get proper jobs and get their own money.' Some of the other kids laughed and said, 'Yeah.'

Miss Foster and Mr Smith took me outside onto the playground and asked if I took the purse and I said no. They said the purse better be back on Miss Foster's desk by the end of the day or else they'd tell my parents. I started to cry. I've never stolen nothing from no-one, not even one of Sophie's lollies she keeps in her top drawer at home. I saw her put them there and I've never even gone into her drawer once.

But Miss Foster's purse wasn't on the desk at the end of the day. Mr Smith called my parents and told them to come to the school. They said I had to give the purse back. I said I didn't have it and they said they would punish me when we got home.

I am in my bed now. I wasn't allowed to have dinner as punishment. And I'm real hungry too cos Ma made that pumpkin soup and hot bread. It smells real good. Ma and Pa B said they was really disappointed in me for doin' it. I still told them I didn't but they wouldn't listen. They said that all Aboriginal people are bad and that I really have to try and be more white and be good. I want to go back to the Home cos Matron Rose never said I was bad or stole somethin' when I didn't. Even when she used to leave all those sixpences and thrupences on her desk and the door open, I never took none of them.

Friday, 2nd July

Miss Foster and Mr Smith took me aside again today and said that they'd expel me from school if I didn't own up to stealin' the purse. I just couldn't stop cryin' cos no-one will believe me, not at school or home. All day I could tell people thought I was the thief and I just hung my head low with shame about somethin' I didn't even do. Oh, I wanna go back to Bomaderry where everyone liked me, even loved me, and I never got blamed for things I didn't do.

Saturday, 3rd July

I cried myself to sleep last night. Everyone's being really cold to me here. I even saw Sophie take her favourite doll and her money box and give it to Ma to mind. She musta thought I would steal them or something. But I don't steal, and especially not from my family too. Who would do somethin' like that? I don't know what I can do to make them believe me.

Sunday, 4th July

Still people think I stole Miss Foster's purse. Ma and Pa B said in the car on the way to the church that we should all pray to God for forgiveness for anythin' we'd done durin' the week that we shouldn't have. Both Sophie and Sam looked at me. Sophie looked sad for me, but Sam had a big ugly grin on his face.

In church I prayed that they would find Miss Foster's purse, that maybe it fell out of her bag or something and then everyone would know that I'm not a thief and they'd say sorry for bein' so mean to me all this time.

I feel really bad but I also said a prayer that I was white. I don't think that if I was white they would have blamed me for stealing the purse. They blamed me cos I was an Aborigine. Everyone thinks Aborigines are bad. I don't even really know what Aboriginal is though. I think it means when people have brown skin, because they always say there's Aboriginal people and there's white people. But no-one really tells us what the difference is. They just say it's better to be white cos Aborigines are dirty, bad and sinful, and that we should try and be white. That's gunna be pretty hard for some Aborigines though cos some, like the kids at Bomaderry, have really dark brown skin and will never even look a bit white. I'm really confused. I don't know what it's better to be any more. I just wished I was treated like everyone else. And that they didn't blame me for the purse.

Monday, 5th July

They still haven't found Miss Foster's purse and it was terrible at school. No-one would talk to me and all the girls were talking behind their hands and looking at me like I was some thief. I'm not. I wanted to cry all day but I just sat there and pretended to read. Toni wouldn't even sit with me either. She must think I took the purse too. Why won't anyone believe me? Miss Foster was really mean to me in class too. She asked me all the hard questions and when I knew the answers she didn't even say I was good or nothin'. She hates me too now.

Tuesday, 6th July

Now Ma and Pa B are saying that they might send me away cos they don't want a daughter that steals. I can't stop cryin'. No-one will listen to me. I didn't take the purse. I don't lie and I don't steal. Why won't they believe me? They can't send me back to the Home. I can't go back like this. It was different when I wanted to go back, by myself, but I can't go back cos they sent me for stealin'. That would be too shameful, eh? Oh, why is this happening to me?

Wednesday, 7th July

You won't believe this but Johnny Jones's mother came to the school today with Miss Foster's purse in her hand. She found it in Johnny's bedroom this mornin' when she went to wake him up. So it means he stole the purse and blamed me. He got in big trouble from Miss Foster and Mr Smith and his mum. Mrs Jones said Johnny wasn't allowed to play football for one month. Nobody said anything to me though, not even sorry for getting me in trouble with Ma and Pa B.

I told Ma B when I got home about Johnny Jones. She just said she knew I wouldn't do somethin' so bad. She let me have an extra big piece of chocolate cake with a glass of milk as a treat. Sophie wasn't here so I couldn't make her jealous but I'm gunna tell her when she gets home that I had a big piece of cake.

Friday, 9th July

It has been such a bad week, what with Miss Foster's purse being stolen and everything and now Dot was really upset when I saw her today. She was almost cryin' and shakin' a bit when I saw her at the shop. She said she met an Aboriginal man on the train near Central when she was runnin' some errands for the Sullivans. This man was runnin' away from his boss. Dot said his name was James and that he'd been in Kinchela Boys' Home and was sent from there to work in the abattoirs. That's where they kill cows for meat. This man loves animals and didn't want to kill them and so he ran away to the city. He thinks he has some family in Sydney and is going to try and find them. Dot said she was worried about this man and that he reminded her of her brother Noel who she hasn't seen for years. That's why she got upset.

Saturday, 10th July

I spent ages doing Sophie's curls today cos I was tryin' to keep myself busy. I was thinking about Dot. I hope she's all right.

Sunday, 11th July

A new hospital opened in the Hives today. It's called the Margaret Reid Orthopaedic Hospital and it's a private hospital for kids recovering from poliomyelitis. That's when kids get a disease that makes them paralysed.

We took a drive past the hospital after church and Sam made some comment about how I should be there cos it's for crippled children and I'm crippled cos I'm an 'abo'. Pa B swung around from the front seat of the car but Sam ducked before Pa's hand could get him. Ma told Sam that good people don't say things like that. There goes that good people, bad people thing again. Now, Sam is white and is supposed to be good, but he says these things that are really bad, so white people can be bad too! Anyway, I reckon Sam is really jealous that Ma and Pa stick up for me now too.

Monday, 12th July

Dot told me some really interesting stuff today about the Wiradjuri tribe of New South Wales. That's the one that she says I come from too. She says the tribe is one of the biggest in the whole of the state and covers a lot of central New South Wales. And people from that tribe speak the Wiradjuri language. Hardly anyone knows it any more cos when the white people came and started living here the government made these laws that said it was a crime for Aborigines to speak their languages and everyone had to learn English. English is what they spoke when they got off them big First Fleet boats, you see.

Tuesday, 13th July

I went to singin' group today but I couldn't really concentrate. I was thinking about all that stuff that Dot told me yesterday. I was thinking that if they didn't have that law then maybe we would be singin' some songs in the Wiradjuri language or even the language of the Ku-ring-gai tribe that used to live here. It doesn't sound right that Aborigines can't live where they want or speak their own languages like white people can, does it?

Friday, 16th July

Dot was at the shop after school today and when she was givin' Mr Johnson the money for the bread and milk, she said here's the 'walang'. Mr Johnson just laughed cos he and Dot get on really well. She's his best customer, I reckon, and he doesn't mind that she's an Aborigine either. Anyway, Dot told me 'walang' is the Wiradjuri word for money. Walang. I'm gunna say that to Sam and he won't know what it is and he'll think I'm calling him some bad name or somethin'.

Sunday, 18th July

I got in trouble in church today cos I couldn't stop laughin'. Neither could Sophie cos when they passed the plate around for the money after communion, I said to Sam, 'Walang! Walang!' He just screwed his face up at me and Sophie and I laughed cos I'm teaching her the words too. We got the giggles real bad cos Sam hates not bein' in control of everythin'. Ma grabbed Sophie and made her sit right on the other side of Pa so we couldn't even look at each other.

Tuesday, 20th July

Today in singin' group we did the song 'Advance Australia Fair' by P D McCormicle. It's a strange song that is meant to make everyone feel proud to be Australian. But the first line is 'Australian sons let us rejoice'. What about the daughters? And it has this line that says 'Our home is girt by sea'. What does girt mean anyway?

Wednesday, 21st July

I asked Ma B at breakfast this morning what 'girt' means and she said 'surrounded', that Australia is surrounded by sea. Now that makes sense, but why don't they just say that instead?

Friday, 23rd July

It was a whole week since I saw Dot and I couldn't wait to learn my new word so I could use it at school. And it's a really good one that I can use on Johnny Jones the next time he bullies me. It's 'gaja' and it means 'getaway!' or 'don't!' He'll never know what it means. Sophie and Toni won't tell him cos they can't stand him either. Can't wait till Monday cos he's sure to give me a hard time again. At least now I've got some ammunition.

Sunday, 25th July

Ma B laughed all the way to church today cos Pa B was talkin' about the Prime Minister, Joseph Lyons, and sayin' that he looked like a koala with his round body, long nose and funny bit of fair hair. Ma said Pa was being very un-Christian like. He said he didn't mind Lyons though cos he was an Irish Catholic like Pa is. And I started thinking, now that I'm baptised into the

Catholic Church, and if I was meant to really be a Burke, then was I supposed to be an Irish Catholic too? How could I be? Dot says I'm an Aborigine, so then I must be an Aboriginal Catholic, eh?

Monday, 26th July

Sure enough Johnny Jones couldn't help himself today and started on at me. This time he said I had nits, but I know I don't cos I have really clean hair all the time and I nearly always wear it in plaits to school. Anyway, he started singin' out, 'Mary has niiiitttts, Mary has niiiitttts.' All the other kids joined in so I just turned around and said really loud, 'GAJA!'

He just stood there and said I was a witch or something and Toni and I walked off around the corner and laughed our heads off cos I told her what it meant. It was so funny. Can't wait to see Dot again to get my new word.

Tuesday, 27th July

Still learning 'Advance Australia Fair'. I told Mrs Elliott what girt meant and she just smiled. I think Joey was impressed too.

Friday, 30th July

I saw Dot today and she told me some really bad stuff about the Home I was in, and other homes like it. I told her how my name used to be Amy Charles and how they changed it to Mary Talence when I got to Bomaderry. She put her arm around me and said they changed my name so my parents wouldn't be able to find me any more. I didn't believe her cos Matron Rose really loved me. I was her favourite. She wouldn't do that cos she knows I wanted to see my mum.

I started to cry and I think Dot felt bad, but she said I needed to know this stuff. She said they changed lots of kids' names so if parents wrote to the home and asked about Amy Charles, the people at the home would say, 'There's no Amy Charles here', because my new name was Mary Talence.

I'm so confused. Why would anyone want to do that? I thought my mum stopped coming to see me cos she didn't love me any more or cos she used to get too upset or maybe even she was dead, like Matron said about the parents of the other kids. But maybe she never came back cos she thought I wasn't there any more, or worse, maybe she thought I was dead.

I am so confused about all of this. Why did this

happen to me and all those other kids in the Home? And why didn't it happen to any of the kids I'm at school with now? Or Sophie and Sam? Who makes these decisions anyway? Dot reckons the government does, but that Aborigines don't have the right to vote for the government, only white people can do that. But white people aren't the ones the government are putting in homes, are they? I really don't understand all this stuff, but I know it's not fair.

I told Ma B I felt sick and couldn't eat dinner cos I really did feel sick from what Dot had told me. Sometimes I wished I'd never been born.

Saturday, 31st July

I'm still confused about why they changed my name. I never thought about it much till yesterday cos I thought it musta happened to everyone.

Sunday, 1 August

Well it's the horses' birthday today. Pa B told us at dinner and then he made this funny horse kinda sound. I don't know why but it made me think about everyone's birthdays and especially my real mum's. I feel bad thinking about her all the time cos the Burkes have been nice to me even though they get angry when I talk about the Home or my other family. They want to pretend that they are my real family even though I look nothin' like them, and Sam treats me meaner than a real brother would.

Tuesday, 3rd August

Today at singin' group Mrs Elliott told us about a competition on that radio station 2WS. They are trying to find the best kid singer for that song about Aeroplane Jelly. I want to go in the competition cos I reckon I can sing it the best. I already knew most of the words anyway, and Mrs Elliott gave us a copy so we could practise.

Wednesday, 4th August

Still practisin' . . . 'I like Aeroplane Jelly, Aeroplane Jelly for me.'

Thursday, 5th August

Well, sometimes I feel like I'm learning more from Dot than from school, but that's all right, as long as I'm learnin' is what I reckon. Today Dot taught me a word meanin' 'to be afraid' or it can mean 'to be ashamed'.

It's 'giyal' and I reckon Johnny Jones should be 'giyal' (afraid and ashamed!) for all he's done to me. I reckon he's gunna get payback one day, eh?

Saturday, 7th August

I sang the jelly song for Pa and Ma B today and they reckon I should win that competition.

Sunday, 8th August

I prayed in church today that I'd win that competition.

Monday, 9th August

Toni invited me to her eleventh birthday party that's gunna be on Saturday. Ma B said I can go. She's even goin' to make me a new dress to wear. I have to get a present for Toni too. I told Ma B I wanted to buy her a diary to write in every day, and she said that was a really good idea.

Tuesday, 10th August

We practised the jelly song today cos the competition people are coming soon to hear us sing. Only about five kids wanna go in it so that's even better.

Wednesday, 11th August

Ma B said she spent all day working on my dress for Toni's birthday. She says Toni's family are real high society and so I have to look and behave my best. My dress is light pink with little roses around the sleeves and the hem. I have to wear white tights and my black patent leather shoes too. Geez I hope they don't get too dirty. Sophie was real sooky about not gettin' a new dress but Ma B said that they'll do something nice on Saturday when I'm at the party. I think that sounds fair, eh?

Thursday, 12th August

Only two sleeps to go until Toni's birthday.

Friday, 13th August

I'm going to bed early cos I'm really excited about going to Toni's birthday tomorrow. It's going to be at her house and I reckon there's going be heaps of really good food there, not like the Vegemite sandwiches we usually have. Pa brought home a really nice diary with a purple cover and Ma helped me wrap it in brown paper with a red bow. It looks real special.

Saturday, 14th August

It was the best birthday party I've ever been to. Toni's house was really big, much bigger than ours. And they had a really big yard, like a paddock almost. I met Toni's brothers and sisters and all her uncles and aunties. There was so many of them I lost count. And the food was really good, tasted really different too. And I know Johnny Jones called Toni 'spaghetti head' before but we really did have spaghetti today and I had to try really hard not to suck it up too loudly or spill anything on my new dress. Everyone loved my new dress too so I was glad that Ma B made it for me. I gave her a big hug when she came to pick me up. I think she liked that.

Tuesday, 17th August

It seems like we sing lots of real Australian songs in singing group cos today we did 'Waltzing Matilda' by Banjo Paterson and Marie Cowan. The boys have to sing the chorus really loud an' it goes like this:

> Who'll come a-waltzing Matilda, my darling?
> Who'll come a-waltzing Matilda with me?
> Waltzing Matilda and leading a water-bag—
> Who'll come a-waltzing Matilda with me?

And when they sung 'my darling' I think Joey was looking at me. Does that mean he likes me?

Wednesday, 18th August

Ma went to town again today to see some friends so that meant that Dot and I could go to Bungaroo for a while after school. When we was there, there was a really cute baby with its mother. The baby was just lying on a rug in the grass and the mother was singing to her some baby songs. I told Dot that we used to get little babies like that one in the Home and all of a sudden Dot got really angry. She reckons she heard how they took babies from their mothers in the hospital. Sometimes the mother never even saw their kids or anythin'. I told her those mothers died like Matron Rose said, and Dot said, 'That's a lie! How could all those mothers just die?' It's funny cos I thought that too when I was in the Home.

I asked Dot why they took those babies from their mothers. She said she didn't know either but officials always told the kids that their parents were dead, or that they didn't have any parents, or that their parents didn't want them. But Dot reckons the mothers always cried and wanted their children back. She said to me: 'If you had a lovely little baby girl or boy and someone took them away from you, wouldn't you want them back?' I said of course, cos I had a doll once and they took that off me and I cried for days and that was just a doll, not a real person or nothin' like that. It also made me think about how they took Marj away from me and then took me away from little John.

All that talk made me wonder if that's what happened with my favourite baby at the Home, little John. I miss him lots, almost like he was like my own. Course, I'm way too young to have a baby. I'm not even married yet and you have to get married first Matron said.

Friday, 20th August

Just got home from school and from seeing Dot and wanted to write this word down before I forgot. 'Muuji' means friend and I'm gunna call Toni 'Muuji' from now on cos she's my true friend.

I think Dot likes seeing me of an afternoon cos she doesn't have a lot of friends round here. She always gives me a hug too, and I really like that.

Saturday, 21st August

It's Ma's birthday tomorrow so I'm gunna write her a song. I'll tell you what it is tomorrow.

Sunday, 22nd August

It's Ma B's birthday today and we had a little party at home for her with all the Burkes and Ma B's sister Joan and brother Sid and all their kids. I made a card for her birthday and sang her her song which she loved lots.

Pa B gave her a book called *Sun Across the Sky* by a woman called Eleanor Dark and Ma was really excited cos it's a new one that just come out. Usually Ma has to borrow books from the library and now she has her own copy. Here's the song I wrote for Ma:

> The prettiest lady is Ma B, Ma B
> She cooks the best cakes and makes the best tea.
> She works really hard to look after Pa
> She's happy to walk when Pa has the car
> And she looks after Sam and Sophie and Me
> The prettiest lady is Ma B, Ma B.

Tuesday, 24th August

Did 'Waltzing Matilda' again and I watched Joey to see if he was lookin' at me when he said 'darling', but he didn't. Boys are so stupid.

Saturday, 28th August

Ma B was really anxious today cos there was a big story in the paper about infantile paralysis. That's the proper name for polio which is making lots of Australian kids really sick, especially in Victoria and Tasmania. Me and Sophie got a bit scared when Ma was worried. What if she thinks we might have it. She reckons that it makes your arms and legs really weak and sometimes it affects your lungs and you have to lie in something they call an iron lung to help you breathe. The iron lung thing is really weird though cos it's outside the body and our lungs are inside so how can that work? I took a really deep breath and hugged Ma B and said we ain't sick at all. I feel sorry for those kids in that iron lung. They can't get up and walk around or anything.

Sunday, 29th August

After church we went to the St Ives Showground for the district agricultural show. There was lots of cows and pigs and goats and stuff and lots of fruit and vegies on display. There was even cake competitions but Ma B didn't put one in there, though I reckon she makes the best cakes in the world.

Tuesday, 31st August

Oh no, another boy song in singin' group today. You'd think all Australians was men. Anyway, we had to sing 'Click Go The Shears' and there was all these really strange words like 'snagger' and 'bare-bellied yoe'. The boys really liked singin' it, even Joey. But I was not really watching him anyway.

Thursday, 2nd September

It's really hard to keep fibbin' to Ma B, but if she just let me talk to Dot then I wouldn't have to, eh? It's not right to tell people who they can and can't talk to, don't you think? I know Ma and Pa B care about me and worry about me, but it's not like Dot is a bad person or anything. She's got a job with a rich white family and she wears really nice clothes and speaks proper and she's really clever. She reads a lot too, and today she had this book that was written by an Aborigine. She says it's the only book by an Aborigine ever published and she got a copy sent to her by a friend who works in a bookshop in Melbourne. It's called *Native Legends* and it was written by a man called David Unaipon in 1929. Geez, that's eight years ago and there's been no more books since then. Dot reckons that most people think Aborigines aren't very smart and so white people just write stuff about them, but Aborigines should be writing their own stories. That makes sense to me cos no-one could write my story except for me. How could anyone else know what I think or feel if they wasn't me? Dot says that you have to walk in someone else's shoes to know what their life is really like. But I have really little feet so I don't reckon anyone would fit in my shoes anyway. When I told her that she just laughed and rubbed my head.

Friday, 3rd September

We have our big sports carnival on Monday so today we practised doin' some races and stuff like that. We did the three-legged race which is hard cos you fall over a lot and Toni just kept laughin'. She's not as fast as me either, but she's my only real friend so I have to have her as my partner. I said I'd practise on the weekend with Sophie so I hope she practises with one of her cousins or brothers.

Saturday, 4th September

I'm glad it was sunny so we could spend the day outside practising for the sport's carnival. Sophie and I practised the three-legged race nearly all day, until I lost my patience cos her legs are shorter than mine and that makes her slower. She kept complainin' that the rag tied around her ankles was hurting. She's so prissy sometimes.

Sunday, 5th September

I prayed in mass today that I win my race tomorrow. That'd shut Johnny Jones up for a while I reckon.

Monday, 6th September

You won't believe this, I won my race and the relay and everyone was pattin' me on the back and sayin', 'That's good, Mary.' Even Mr Smith and Miss Foster came and said how good I'd done. I was real fast and real popular. It's amazin' how when you are good at somethin' like sports, people forget that you are different to them. Does that mean I have to be really good at somethin' every day just so people are nice to me? I hope not. That'd be too much work I reckon. And who can be good at somethin' every day anyway? OK, so Sam might say he could (but he'd be wrong!).

Tuesday, 7th September

I didn't go to singin' group today cos my legs are a bit sore from yesterday, but everyone was still talking about me winnin' so don't really care how sore I feel.

Wednesday, 8th September

Toni said today was the day that the Virgin Mary was born so it was her birthday. Her family are really religious too, cos the Pope is in Rome in Italy where all Toni's family come from. I didn't know what a virgin was and neither did Toni. So I asked Ma B when I came home. She went all funny and said she was too busy to tell me but would tell me another time. I wonder if I should be called 'Virgin Mary' too cos my name's Mary.

Thursday, 9th September

It was a really nice spring day today and the flowers are starting to bloom, the birds was singin' and the sun was shinin'. And as I said I loved to feel the sun on my face, Dot said that I should call it 'yirra' cos that means sun in the Wiradjuri language. 'Yirra'. I like that word.

Saturday, 11th September

As usual, I did Sophie's hair for her today and it looked really pretty and she was really happy with me. I think I'd do it anyway, even if I didn't have to, cos Sophie and I are becoming more like real sisters now.

Monday, 13th September

I did Sophie's hair again today, because she said if I did it she'd let me use her tambourine and read some of her books. That was fair I thought. It reminds me of when I was in Bomaderry though and did all the little girls' hair. That makes me a little sad cos I still miss everyone.

Tuesday, 14th September

You guessed it, another boy song. This time it was called 'Our Don Bradman'. It was about that famous cricket player that hit all the runs for Australia.

Wednesday, 15th September

I knew it couldn't last too long with Johnny Jones keeping to himself. Today we was in class and I knew the answer to a really hard math question. Miss Foster was real proud of me. I reckon cos no-one else could do the sum on the board for her. Anyway, just as I went to sit down, Johnny Jones looked at me, screwed his face up and said, 'Just cos you could do that doesn't mean anythin'. Don't forget where you come from. You're an abo from an orphanage with no parents. That means you're nothing.' That really hurt me cos I thought I was doing good at school. He thinks he's so important cos his dad is a doctor and all the kids from St Ives Public go to him and most of the teachers too. I'd rather die than go to Johnny Jones's dad. And I'm not nothin'!

Thursday, 16th September

I told Ma B at breakfast today what Johnny Jones said yesterday. I didn't want to be a dobber or anythin' like that, but it made me angry and hurt my feelin's. She said that it's because it's not a Christian school and that Johnny Jones is not a very nice person—like I didn't already know that! Pa B said that when Sophie and I go to high school we can get one of the private buses and go to a Catholic school in Pymble. Pa really wanted us to go to a Catholic school now but there's none in St Ives. Ma B wants us to go to an all-girls school anyway, which I'm happy about cos then there'll be no Johnny Joneses. You know, I realised there was no bullyin' in the Home, only here where I'm the only different kid.

Friday, 17th September

There was lots of drillin' and hammerin' at school today. It was really noisy and we could hardly hear Miss Foster. They're buildin' new classrooms and stuff so there's more space.

Saturday, 18th September

I sang that Don Bradman song for Pa and he loved it. He made me sing it for his mates too when they was havin' a beer. That all clapped me and wanted me to teach it to them.

Sunday, 19th September

Today the church had a picnic that all the families went to. It was a lot of fun. I think Ma and Pa B had a good time. I know I did. We bobbed for apples and I've never done that before. There was a long piece of string with lots of apples tied and hanging off it, and all us kids had to stand under it and try and eat the apple without using our hands. It was so hard cos you have to try and rest the apple between your shoulder and your chin and then bite it. And I kept laughing cos everyone looked so funny. Some boy won and he ate the whole apple. I reckon he musta done it before though.

Next they had a tug-of-war for the men. All the men from Pymble, Gordon, Roseville and St Ives stood on one side, and all the men from Wahroonga, Lindfield, Killara and Waitara stood on the other side. When they made sure there was the same amount of men on each side they started the tug-of-war. All the men was concentrating so hard, frownin' and groanin' and had their sleeves rolled up. They was diggin' their feet into the ground and after about five minutes it was all over. Pa B's side won, thank goodness, cos you know how he likes to win at sport, eh?

But, that wasn't the best bit. They also had a tug-of-war for the women and Ma B even went in it. She's such a lady she looked a little out of place. About as out of place as I looked being the only brown kid there, eh, but no-one said nothin' to me about that, so it was a peaceful day.

There was running races too and Pa B took his shoes off and raced and came second. But on the way home he said the man who won cheated cos he started before the gun went off. Sophie and I just rolled our eyes at each other and smiled. We know how Pa B likes to be the winner always, and is a bit of a show off. Ma reckons all men are like that though cos they want to impress their women.

Someone even had a camera that makes movies but I don't know who it was or if we'll ever get to see it. I hope so.

Monday, 20th September

Today we had a double day of sport cos Miss Foster was sick and we had to sit in Mr Yates's class with the Grade Fours, so Sophie was there too. We had races and relays and everyone wanted me on their team cos I ran so fast at the sports carnival. It was the first time I felt like people might like me and want to be around me, but Johnny Jones ruined it. He said that 'Abos only run fast cos they chase kangaroos for dinner', and everyone laughed. It made me so mad and ashamed again that I had to run really fast and beat Johnny Jones's team. I was the fastest in the whole class I think, and I wasn't even thinking about kangaroos.

Tuesday, 21st September

I'm still tired from all the running yesterday. I went to singin' group but I think even my voice was too tired to sing properly. I couldn't even be bothered watchin' to see what Joey was doing.

Saturday, 25th September

There was a lot of excitement in the house today as Pa B and some of his mates listened to the wireless and the big football match was on. It was the grand final of the Victorian Football League that Pa B calls VFL and Geelong was playin' Collingwood. It's really a Victorian game but Pa just likes every kind of sport—especially the Rugby Union they play here in New South Wales.

All through the game the men was cheerin' and punchin' the air and stuff. Ma B just kept sighin' and shakin' her head. Geelong musta been Pa's team cos they won and I saw some of the other men give him some money at the end. He rubbed his hands together and said, 'I'll be takin' your money again soon, fellas.'

Monday, 27th September

Someone brought this new drink to school today called Coca-Cola. I didn't get a taste of it, but it's supposed to be this really special drink from America. It looks really dark brown and has bubbles in it. Johnny Jones reckoned he'd had some before, but I reckon he was lying. Toni said she'd ask her parents if she could have some and she promised to share it with me. That's what Marj would've done too if she had it. I miss her sometimes.

Tuesday, 28th September

Someone came to our singin' group today and listened to us sing the jelly song. Only the ones who wanted to be in the competition had to sing, but the whole group stayed and listened for support—so that was good. I think I did a good job. I really hope I win.

Saturday, 2nd October

We all sat round the wireless tonight to listen to the Dad and Dave show. It's a program about life in the country in Australia and Ma and Pa B love it. We all have to sit there and sing 'Along the Road to Gundagai'. Sophie and I giggle most of the time though cos we think it's a funny song. I sing real loud too cos I've got the best voice.

Sunday, 3rd October

The car wouldn't start today so we all had to sit around the wireless again and listen to the Australian Broadcasting Commission. Pa calls it the ABC. We listened to the church service on the wireless. I don't know where it was coming from though, and it was really hard to sit still and listen properly. It was nowhere near as fun as listening to the Dad and Dave show. And I bet we all looked real funny sitting there in our Sunday best around the wireless.

Tuesday, 5th October

Well, I'm really sad and angry cos I practised so hard for that song competition and Mrs Elliott told us today that some girl called Joy King won, and that the jelly people are gunna use her singin' the song on radio commercials and everything. I'm so mad about that cos I reckon I shoulda won, but Mrs Elliott just said, 'Next time.' I'm never eatin' that jelly again.

Wednesday, 6th October

Johnny Jones heard I didn't win the competition and started making fun of me cos he knew how much I wanted to win. He said, 'As if they'd let an abo win. Abos don't eat jelly anyway. They eat grubs.' I just ran to the toilets and cried and Toni followed me and gave me a big hug and said Johnny Jones is stupid. That made me feel a bit better.

Sunday, 10th October

In church today Father Frank was going on about people not being pure and I thought, Oh no, here he goes on about Aborigines or somethin', but he was talkin' about some painter man in Melbourne called Norman Lindsay who paints pictures of women in the nude. That means with no clothes on. He said it makes women out to be objects of men and that's not right. But I think he would have to say that cos he's not even allowed to have a wife let alone look at paintings of naked women.

Thursday, 14th October

We did some dancin' lessons at school today, the barn dance and Pride of Erin. We also did something called the Lambeth Walk. I even had to be a partner with Johnny Jones but we didn't hold hands or nothin'. He reckons he wasn't holdin' my hand, but I wasn't holdin' his. He has germs.

Friday, 15th October

When I met Dot in the shop today I showed her the dance steps I learned yesterday and we had a good laugh. As we walked home she told me that Aboriginal people had dancin' too and music. She said it was called a corroboree, and it's like bush opera cos it tells stories through singin' and dancin'. She reckons the first recorded corroborees in Sydney used to happen down at Bennelong Point on the other side of the harbour where the ferries go. The dances and songs they do at the corroboree mean somethin' and usually tell stories of hunting, fighting and relationships. They make music by singin' and the men play something called a didjeridoo and the woman play clap-sticks. I want to go to a corroboree one day and learn them dances. But Dot doesn't know where they happen now, cos it was like with the language stuff: Aborigines aren't allowed to practise their dancin' any more either.

Saturday, 16th October

I'm thinking about that Aboriginal dancin' that Dot talked about. I wonder what it looks like and what the music sounds like. I wonder if my real mum and dad can dance like that.

Tuesday, 19th October

Singin' group practice wasn't on today cos Mrs Elliott was sick, so I went to the shop in case Dot was there and she was so we walked for a little while. She told me that an Aboriginal man called William Cooper has collected one thousand, eight hundred and fourteen names on some paper from Aborigines from all over Australia to get an Aborigine in the parliament. He wants to give the list of names, called signatures, to the King, but the government won't let him. Dot says when you collect signatures like that on a piece of paper it's called a petition. Imagine having over a thousand, almost two thousand, friends too. That Mr Cooper must be a really nice man to have so many friends. I tried to make a list of people I know and my petition wouldn't have many names at all.

Wednesday, 20th October

I started a petition today at school to get rid of Johnny Jones. I already have me, Sophie, Toni and some of Sophie's friends cos he picks on them too. Might be hard to get almost two thousand names like Mr Cooper did, though.

Saturday, 23rd October

Big news today on the radio and in the paper and that's that Japan invaded China and the Australian unions people want the government to stop tradin' with Japan to punish them. Pa B says we should, 'Show them Japs a thing or two', but Ma B says he shouldn't call 'em 'Japs'. That's not very Christian-like. Sam repeated what Pa B said and Ma was furious, and said, 'No wonder that boy says such terrible things to his sisters!' I don't think Ma and Pa was talking at dinner time.

Sunday, 24th October

After mass today Pa B drove us to the Sanitorium in Wahroonga to see some friends of Ma's. There was a big park with loads of people just havin' fun. There was kids running and playing, women sitting around knitting things (don't know why cos it's summer nearly), there was a band playing music like to march to, and they was even dishing out milk for people to drink.

I liked it cos it was really open, with nice grass to run on and a few little trees. The only thing missing was that there was no Aboriginal kids or families. It's still strange how there's no Aboriginal people around.

Tuesday, 26th October

Today in singin' group we did a song called 'Botany Bay'. It's a song that the English people used to sing when they first come to Australia. It's got a funny chorus and it sounds like everyone was a convict that came here. That'd make them bad people, eh? But they was all white people though, so there you go. White people can be bad too. I knew it! I like the chorus though. It goes like this:

Singing too-ral li-ooral li-ad-dity
Singing too-ral li-ooral li-ay;
Singing too-ral li-ooral li-ad-dity
And we're bound for Botany Bay.

Friday, 29th October

Dot was at the shop today after school and walked to my street with me. She gave me and Sophie a lolly each which we ate right away. I was sayin' to her that everythin's much fancier here than in the Home. You know, things are much better: nicer clothes and shoes, better food and not so many chores. Dot looked at me and said, 'Do you feel loved though?' and I didn't know what to say. I feel like the Burkes really care about me, but I don't know if they love me like my own parents.

I'm a bit sad tonight. Maybe Dot's right. Maybe love and family are much more important than shoes and lollies and stuff like that. Why can't I have both though? Sophie and Sam and the other kids round here have it.

Saturday, 30th October

Pa B took Sam fishing today and me and Ma and Sophie all waited anxiously for them to come home so we could cook a yummy dinner. Fish and a salad and some of Ma B's home-made bread. But the boys came home and when they got the fish out there was some dark stuff on it. Us three girls all looked at each other and we knew that they had bought the fish from the shop and it was wrapped in newspaper. They was too embarrassed to say they didn't catch anythin'. Ma B just winked at us girls and we knew not to say anything cos Pa B would be really upset if he thought we knew that he didn't catch anythin'. I really wanted to say that my real dad always came home with lots of fish from the river. He couldn't afford to buy fish so he had to catch our dinner for us. I didn't say anything though. That would have made Pa B feel bad about himself and it woulda made him angry cos I'm not supposed to talk about my Aboriginal family any more.

Tuesday, 2nd November

You won't believe what we did this afternoon in school. We listened on the radio to this big horse race called the Melbourne Cup. It's held in Melbourne at the Flemington Racecourse. They've been havin' it every year since 1861. I've never heard of it before cos we didn't listen to it in the Home and of course Johnny Jones couldn't help but say somethin' about that, eh? I just ignored him though and so did most of the other kids who are gettin' sick and tired of him, I reckon. Miss Foster says all the ladies get dressed up for the race and wear hats and some of the teachers even had hats with flowers on today too. You can put money on what horse will win. I think Pa B won some money cos he was in a really good mood tonight when he came home with a coupla bottles of beer and it's not even Saturday or Sunday. The horse that won was called The Trump.

Monday, 8th November

The new part of the school opened today and it's great. You should see the new classrooms. There's so much room. There's new rows of seats to sit on in the playground and a brand new flagpole. Before, Mr Smith would get two of the older boys to stand out the front of assembly and hold the flag up for everyone to see. The school's now got two storeys too which makes it one of the biggest buildin's in the whole area. Johnny Jones said the Governor General of Australia, Lord Gowrie, was gunna come and say a speech but he never did. Mr Smith just made a speech and everyone clapped and cheered and stuff.

Tuesday, 9th November

We started getting ready for the Christmas pageant today and learnt 'Silent Night' in singin' group. I really like that hymn.

Thursday, 11th November—Armistice Day

At 11 am today we all had to have a minute's silence to remember the day when Germany signed an agreement with the Allies to stop The Great War. The agreement was called an Armistice which means an agreement to stop the war just while they talk about how to have peace. It was signed in France on 11/11/1918 at 11 am.

Friday, 12th November

I haven't seen Dot for ages. I wonder what she's doin'. I hope she's not angry with me or something. Maybe she's gone on a holiday, but she never told me.

Tuesday, 16th November

Today we practised 'O Come All Ye Faithful' for the Christmas pageant.

Wednesday, 17th November

Remember when I first started at St Ives and Miss Foster said that Aborigines were a dying race and I didn't really know what that meant, well today's paper said something about it. Professor Wood Jones from the University of Melbourne reckons that dying race stuff is 'humbug with which the white man has always gilded his extermination of native races'. I'm not sure what that means, but he reckons that 'The Aborigines were never a dying race until we started to make them die'. He also says that there was almost 300 000 Aborigines when the boats first arrived here from England and now there's about 60 000. That's real scary, I reckon. I hope my family aren't part of that big number of people that have died.

Thursday, 18th November

We had bananas today. What a treat. Ma got them from Mr Johnson's shop, but he hardly ever has them cos they go bad when they transport them down here from up the coast. He reckons they need to be refrigerated when being brought to Sydney and they can't do that, so he only gets them sometimes. They was so sweet, love 'em love 'em. Only bad thing was when I was eatin' my banana Sam said I looked like a monkey and that 'abos grew from monkeys', and that's why we was brown. His theory scared me a bit.

Friday, 19th November

Ma offered me a banana today but I said no. I didn't want Sam to say anything. I was mad then cos I really wanted a banana.

Saturday, 20th November

It's Pa's birthday next week so Ma B took me and Sophie and Sam into town to shop for presents. We went to Anthony Horderns Department Store and there was so many people I couldn't stand it. I was glad to come home. Anyway, Ma bought a book for me and Sophie to give him. It's a book by Helen Simpson called *Under Capricorn*. Sam's gunna give him boring socks cos he couldn't be bothered thinking too hard what to get. And Ma got him a new hat, which he'll really like.

Sunday, 21st November

Everyone was talking at church today about what was in the newspaper about how a man called Hubert Opperman rode his bike right across Australia from Fremantle in to Sydney. That's 2751 miles in just eighteen days and eighteen hours. Wow, he must have really strong legs, eh? And must peddle really fast. I couldn't do that. I'm flat out riding Sam's bike round the block when he's not home. *The Age* newspaper said that the man fell asleep a coupla times and fell off his bike. No wonder. He musta wanted to lie down to sleep. There was lotsa people and cars watching him arrive in Sydney late last night right along George Street and near the General Post Office.

Monday, 22nd November

I looked for Dot today, but she still wasn't at the shop. I wonder where she is, eh? I hope she's all right. Everyone I like seems to go away, or gets taken away, or I get sent away. Is it me?

Tuesday, 23rd November

Oh, today's Christmas carol was a hard one. We did 'Good King Wenceslas' and couldn't say that properly, wen-ces-las. It's pretty hard when you're tryin' to sing and keep in tune and everythin'. I'll have to practise more before the pageant.

Wednesday, 24th November

It's Pa B's birthday today. I don't know how old he is, but

it must be old cos he's over thirty I think. Ma made him a pavlova for a special treat and he really loved it. It's like a meringue cake with lots of cream and strawberries and passionfruit on top. When Ma B was making it she told us that it was named after that famous ballerina called Anna Pavlova who came to Australia in the 1920s. A man from Western Australia invented the pavlova cos he really liked her.

Sam is such a party pooper though that he wouldn't have the pavlova that Ma B made. He wanted some Peters ice-cream instead. Ma let him have it cos she didn't want to upset Pa on his birthday. Sam is so insensitive sometimes—I mean most times.

Sunday, 28th November

We went to church today, of course. It was the First Sunday of Advent. I'm not sure what that is but it's a special day in the church.

Monday, 29th November

I was really upset at school today cos they started practising for the Christmas play. I want to go in it and sing 'Silent Night' and the other Christmas carols we've been learning. And I wanted to be Mary cos my name is Mary anyway. But Mrs Elliott said that I couldn't be Mary, cos Mary the Mother of Jesus was white and I'm not white. I said that maybe Mary hadn't been in the sun enough to get a tan. Mrs Elliott got really, really annoyed with me and said I was being cheeky and that maybe I shouldn't be in the play at all. That made me shut up, eh? So, she let me be an angel, but I wasn't gunna ask her if there were brown angels in case she really did stop me being in the play. Toni's gunna be Mary so I'm happy about that, but Johnny Jones said behind his hand that she's a 'wog' and shouldn't be Mary. He wanted to be Joseph and thank goodness Miss Foster picked someone else. Johnny Jones couldn't even be one of the three wise men. He had to be the donkey and I couldn't stop laughing.

Tuesday, 30th November

Still practising 'Good King Wenceslas' . . .

Thursday, 2nd December

Toni and I was talkin' today about how people call her a 'wog', and we was trying to figure out what it meant. We reckon it means 'Well Organised Girl' and so she said she's happy to be a WOG then cos she's better than the other girls. She reckons that ABO must mean 'A Beautiful Ornament' and I didn't mind that too much either, but I know that's not what Johnny Jones means when he says it, eh?

Saturday, 4th December

We went to the markets today in Quay Street, Haymarket. That meant driving over the Harbour Bridge and that was really exciting. The markets smelt really different cos there was so many different foods there like fish and chickens—real live chickens that they wrapped in newspaper and let their heads poke out the top. There was also rabbits and I wondered how Ma B felt cos she has that fur coat and all. There was fruit and vegetables and even some bananas all the way from Mullumbimby. They smelt real good and I wanted some but didn't want to ask in case of you-know-who saying somethin'. There was hardly any horse and carts but lots of cars lined up right along the side of the market building.

Ma and Pa B got in a bit of an argument. Apparently there had been this big research done on the fruit industry this year and the report didn't say anything about how women were really important in pickin', gradin' and packin' the fruit. Pa said it really didn't matter and Ma B was furious and said he didn't appreciate women at all. I'm not quite sure what that means but by the tone of Ma B's voice, I think it might mean he doesn't like women very much, which is pretty silly cos he's married to one, eh?

I didn't even want to ask if that report talked about how Aborigines pick lots of the fruit. I know they do that cos Dot told me that she was going to pick cherries in Young one day with all the other Aborigines over there. Hey, maybe that's where's she's gone. I hope I see her soon.

Sunday, 5th December

After mass today we went to Manly Beach again. It was a real sunny day so we all went swimming—well not Ma B cos she says she can't go in the sun cos she's too fair. I loved it, splashing around in the surf and the whitewash. Sophie and I built a really big sandcastle. We was there for ages, but Ma B said I shouldn't spend too much time in the sun cos I'll get too dark. She didn't want that. I asked her why and she said, 'Life will be hard enough for you, my dear, without having really dark skin.' But that didn't make sense to me. Why should life be harder just cos I got darker skin than Sophie or Sam or Ma B?

When we got home, I heard Ma talking to Pa saying that I looked like a real Aborigine today cos I went real dark in the sun, and that I shouldn't go in the sun any more. That didn't make sense to me. There was all these people lying on the beach today and it looked like they was tryin' to get brown like me, like they was tryin' to become a dark person. I wonder if it will make their life harder too, the darker they get. This skin colour thing doesn't make any sense to me at all. You're really lucky that you're just a diary and not a person.

Monday, 6th December

At school today Johnny Jones was calling me 'abo' and 'boong' and telling me I should go back to my own people with the spears and the boomerangs. He said I really was a 'boong' and pointed to my brown arms. Ma was right! Life does get harder the darker you are. He was even meaner today cos I went to the beach yesterday. Oh, maybe I really should stay out of the sun for a while and it won't be so bad at school then.

Tuesday, 7th December

I tried to sit in the shade today but Toni was bugging me to go and play in the sun with her cos it was a really beautiful day. I think she got angry with me but I'm too embarrassed to tell her why.

Wednesday, 8th December

Toni went and played with other girls today cos I wouldn't go in the sun. I don't think she wants to be my friend any more. That means I can't have friends if I'm dark and I can't have friends if I try not to be dark. I can't win. I hate school, and I hate this skin colour thing that makes people act funny towards me.

Thursday, 9th December

I wrote Toni a note and explained why I can't sit in the sun. I gave her the note in class and at lunchtime she sat with me and didn't say nothing about it. She's a really good friend, eh?

Friday, 10th December

Sophie and I did disgusting things today. We collected Christmas beetles and put them in Sam's shoes outside the door. He didn't see them and put his foot in there and started yellin' when he heard the crunch. Poor beetles, but it was worth seein' him angry and embarrassed cos he acted like a little boy, no, a little girl. We were both hiding in the backyard and he couldn't see us laughing at him.

Saturday, 11th December

At dinner tonight Sophie asked me if I'd seen any Christmas beetles this year. I said, 'Yeah, and they looked real funny cos they was wearin' these really big shoes!' Well, we both fell about laughing and Sam got up and stormed out. Ma and Pa B didn't know what was going on but they laughed cos we was laughing so hard. It feels good to get Sam back cos he's always picking on me and pulling Sophie's hair.

Sunday, 12th December

Sophie and Ma and I sang 'The Twelve Days of Christmas' in the car all the way to church and back. 'And a partridge in a pear tree . . .'

Monday, 13th December

We went to the shops today with Ma B to send cards for Christmas to her family in Melbourne. She said there was special Christmas stamps for the first time this year that was designed in Austria. That's a little country on the other side of the world near Germany. She couldn't get any of those stamps cos she needed Australian stamps but she said maybe next year Australia will have Christmas stamps too and that would make the envelopes look pretty.

Tuesday, 14th December

Today we learnt 'We Three Kings'. Still doing the boy songs, eh? The Christmas pageant isn't far now.

Wednesday, 15th December

It's only ten days till Christmas and I've been thinking about the kids at Bomaderry and rememberin' all the Christmases I had there. I can't really remember the ones with my real parents. At the Home, there'd be all these presents donated to the kids and they was great. About a week before Christmas day we'd go to the people's homes from the church or they'd hire a hall and we'd have a big Christmas party. I reckon they'll probably do it again this year. I wonder if anyone will think of me.

Friday, 17th December

School finished today, so we're on holidays for over a month. Freedom at last! No more teachers, no more books, no more Johnny Jones's dirty looks. Yay!

Still no sign of Dot, and I wanted to tell her about all the carols I've been learning and everything I've been up to.

Saturday, 18th December

Did you know that there is even a place called Christmas Island? Imagine that? Christmas all year round. Wouldn't that be fun, eh? It is a colony of Australia too so maybe one day I can go there.

Sunday, 19th December

I was thinking about Bomaderry in mass today. I wonder if anyone from St Ives used to send gifts for the kids in the Home. I can't imagine that Ma and Pa B would, cos I don't think they really like Aborigines mostly, except for me of course, but they especially don't like Baptists, eh? I think the Burkes might even love me a bit cos I don't give them too much trouble, not like Sam being nasty all the time and Sophie whining about everythin'.

Monday, 20th December

Sophie and I went to Mr Johnson's shop today cos we didn't have any Christmas presents for anyone yet. We decided to pool our money and buy gifts for Ma and Pa

B from the two of us, so we could get somethin' better. Mr Johnson had a whole table of special Christmas things we could look at. We bought Ma B a book by a woman called Ernestine Hill called *The Great Australian Loneliness*. Mr Johnson said that she'd really like it cos he knows how much she likes to read and it's a book that only came out this year so she's probably never seen it yet either.

We got Pa B a really nice navy blue tie and we even bought Sam somethin'—a whole bag of lollies. We both hope all his teeth fall out so he looks really ugly and silly. Hope Ma and Pa B don't mind that we bought him sweets, but it is Christmas and we really couldn't afford to buy anything else. Mr Johnson even gave us a couple of lollies each for bein' such good customers.

Tuesday, 21st December

Final rehearsal today. I think it's gunna be great.

Wednesday, 22nd December

We had the Christmas play at the school hall tonight and it was so much fun. I even forgot about not being allowed to play Mary. Toni did a really good job. Maybe Mary was Italian anyway. Toni reckons the name Mary comes from Maria and that it's probably an Italian or Greek name. Ma B did a good job on my costume and my wings only fell off once.

I sung real loud too, and I could tell that Ma and Pa B could hear me cos they was smiling at me the whole time. They said I made them proud.

PS: I didn't get 'Wenceslas' wrong either!

Thursday, 23rd December

Today Sophie and I stayed in our room and made Christmas cards for Ma and Pa B and for each other. I made one for Dot too but I had to hide it so Sophie didn't see in case she told Ma and Pa B. We even made one for Sam, but we both said he didn't deserve it. We wrote a good message inside which Sophie and I came up with all by ourselves: To our favourite brother, Sam, from the world's greatest sisters. Love and kisses, Sophie and Mary.

He'll hate that cos he hates mushy stuff. Says it's all girly. I bet he doesn't give us one either.

Friday, 24th December

Today we was all very busy gettin' ready for Christmas Day tomorrow. We had to decorate the Christmas tree that Pa B brought home yesterday. Ma B also had some

Christmas bush growin' in the garden and she let me cut some for her to put in vases around the house. She also got some Christmas bell flowers from next door which have pretty yellow tips.

We had to cook the plum pudding for tomorrow cos it had to steam for six hours. We wouldn't have had time to do it in the morning after mass. We used to have pudding at Bomaderry too and Matron would put sixpence or threepence in there and always made sure I got it.

I don't know why it's called plum pudding cos Ma B used apples, raisins, spices and lard but no plums. She didn't know why it was called plum pudding either. She also hid a penny in it for good luck. Hope I get it.

I am so tired now cos we really did a lot today. I can't wait to go to bed and wake up in the morning. Ma and Pa B are still listening to Christmas carols on the radio and the man on the radio called Norman Banks said he could see a woman sittin' in her window listenin' to the Christmas carols by candlelight. Ma said that next year she'd light candles too and we'd all have carols by candlelight.

I can't even think of next year. I just wanna wake up and it's morning.

Saturday, 25th December

Merry Christmas!!!! You won't believe what I got for Christmas—a new guitar. It's just a kid's one but it's great. It was wrapped up with a big bow on it, a big red one, and a card to me from Santa. I reckon the Burkes must have gone and bought it for me which is really nice. I think they really love me and want me to be happy even though I'm not their real daughter.

Sophie got a new doll with lots of doll's clothes as well as some really nice books, including an Australian book for children called *Dot and the Kangaroo* by Ethel Pedley. It was written in 1899, so that makes it really old. I wonder where Ma and Pa B got it from. Anyway, Sophie really likes to read and spent the whole day reading the stories out loud. No-one was listenin' though cos she got real boring after a while.

Sam is really spoilt and got a brand new shiny black bike with big handle bars and he rode up and down the street lots today. He wouldn't give me a dink though cos he's just a meany. Sophie and I said we'd sneak a ride when he's not around one day. I reckon we should let the air out of the tyres too. We help each other out that way. We both asked who would get his old bike. He said he'd rather give it away than let us have it. He is really not a nice boy at all.

Pa bought Ma this really pretty bathing costume to take on holidays with us. It's like shorts on the bottom and the top tied around Ma's neck, and it had dots all over it. Ma said it was very daring and she didn't know if she could wear it or not. Pa said she looked really pretty and he had a bit of a sparkle in his eye.

I played my new guitar for the family and everyone clapped and I sang 'Jingle Bells, Jingle Bells' and 'Silent Night'. I love my guitar.

We ate so much food, yummy turkey and ham and big crispy potatoes and some pumpkin. And we had pudding for dessert. Yummo.

I even thought about my real parents today and what they would be doing. And all my brothers and sisters. I wonder if they had lots of food and presents, and if they are thinking of me. I missed my mum lots today cos she would've given us all big kisses and hugs even if we had no presents. And she wouldn't have made me go to mass at midnight last night. I nearly fell asleep I was so tired.

Sunday, 26th December—Boxing Day

Today we had Boxing Day, which I've never had before and Sam started making fun of me again cos I'm doin' all this new stuff. He said I was primitive and Pa B made him go and wash the car for being so horrible. He is always remindin' me that it's shameful to be an Aborigine. Sophie and I laughed anyway and said it was worth him making fun of me just to see him get punished. Anyway, Ma told me that Boxing Day is always the day after Christmas and it's meant to be when all the empty Christmas boxes are filled with leftovers or small presents to give to poor people. I reckon some of these boxes could be given to my real family cos we was real poor when I lived with them and we didn't have no fancy leftovers or nothing to wrap up.

Monday, 27th December

We're going on our holiday today. I'm gunna pack you in my bag, but I mightn't be able to write every-day cos I don't know what it's like where we're going. I'm excited though cos I've never been on a holiday before but I'm not gunna say that out loud cos you-know-who-will have a good time making fun of me, eh? See ya again soon, OK?

Thursday, 30th December

We're in Coolangatta which is in southern Queensland. It's north of Sydney and St Ives and it took us forever to get here. They say this place is warmer in winter and cooler in summer but it's pretty hot now. We're gunna go surfing, swimming, boating, hiking and play tennis. I can't wait. Ma B is makin' me wear a hat though and always tellin' me to cover up but it's so nice in the sun. I don't care if I change colour.

Saturday, 1st January 1938

It's the first day of 1938 and we had fireworks last night on the beach and people was partying everywhere. Lotsa music and laughin'. Ma and Pa even let us stay up till midnight.

And boy did Pa look sick today. He reckons he ate some bad oysters but Ma said he washed them down with too much beer. I'm never gunna drink beer cos I'd hate to be as sick as Pa looked.

Saturday, 8th January

Wow, everyone is talkin' about Australia's big 150th birthday party on 26th January. Ma and Pa B are organising with their friends to take us kids into the city for all the excitement. There's gunna be marching bands and boats and music and stuff. Sounds really exciting.

Sunday, 9th January

We got home this afternoon and Ma asked me to get some milk from Mr Johnson's store. Dot was there. I hadn't seen her for ages. When I went to give her a hug she pushed me away and it felt like my heart broke into a million pieces. I tried again and she said, 'No, I can't talk to you any more. Go away. I'll get in trouble.' I was all confused and I started to cry cos I didn't know what I'd done. Dot started to cry too and she took me to the corner of the store and bent down to talk to me. She said that someone had told the Burkes about me and Dot and Ma B rang the Sullivans. Dot got in real big trouble. The Sullivans said that if she kept talking to me they'd fire her and she wouldn't get another job. So she's been stayin' away. She said she loved me but that she had to go. She gave me a hug and left me standin' there crying. Mr Johnson gave me a handkerchief and a lolly but it didn't help.

I ran really fast after Dot and grabbed her around the waist and said she couldn't do that to me, cos everyone was always leaving me or going away, like my real mum, and Marj and Matron and little Johnny, and now her. I told her she was the only one who hugged me proper, like a real sister, and she couldn't leave me. I couldn't stop cryin' and then Dot couldn't stop cryin' neither.

Finally she said she would work something out and to meet her at Bungaroo tomorrow about 3 pm. I said I would.

I'm so confused right now though. Why didn't the Burkes say something to me about Dot? No-one said nothin'. Maybe they think cos I'm a kid it ain't my fault or something. That'd be a first. Maybe after everyone blamin' me for Miss Foster's purse, they thought they'd blame someone else. Well, that's all right with me then, cos I had lots of blame already I reckon.

I'll tell Ma B that I'm meeting Toni at Bungaroo tomorrow and then go see Dot.

Monday, 10th January

When I met Dot at Bungaroo she made a little picnic and said she was thinkin' of me on Christmas Day and she gave me a little present and a lovely card. The present was four ribbons: one red, one pink, one green and one really special blue one. Dark blue like the ocean. I put it in my hair straight away. I gave her the

Christmas card I made for her and she loved it, even though it was a bit late.

We sat there for about an hour and Dot told me about this big thing happening in the city on 26th January. It's for Aborigines only, and she wanted to take me with her, but now that she has been in trouble with the Sullivans she said she can't. I just listened anyway.

Dot reckons this Aborigines' thing is much more important that the parade the Burkes are going too. It's a meeting about citizenship rights for Aborigines, she said, because Aborigines weren't allowed to vote in the elections, or drink alcohol, or go to swimming pools. Dot even said that in some parts of New South Wales Aboriginal people had to sit in special areas of picture theatres that were roped off. I didn't think that was very nice. I mean we are all people, aren't we? We ain't animals or something. I got really sad when she told me that a lot of white parents didn't want their kids to go to school with Aboriginal kids either. That's terrible, I didn't know that. I wonder if the Burke's know this stuff or not. Maybe I should tell them.

When I got home no-one even asked how Toni was so I didn't say nothing.

Tuesday, 11th January

There was a poster in Mr Johnson's shop today advertising what's gunna happen on 26th January. This poster had the Sydney Harbour Bridge and fireworks in the background with lots of people in those funny clothes, you know the people that got off the First Fleet had and how they wore them funny hats and white pants and long black boots.

Wednesday, 12th January

All the Burkes are gettin' real excited about the big parade that's gunna be in the city on 26th January. I pretend I'm excited too but they don't know that I'm planning to go with Dot. Dot doesn't know either yet. They'd be really angry if they thought I was spendin' time with her and that I planned to go to that big Aboriginal Conference at the Australian Hall in the city.

Thursday, 13th January

There is a real excitement around Sydney, with posters and everything about Australia Day on 26th January. I saw a poster today with a man's face over the Harbour Bridge. It looked like he was sayin' 'Come here' with his finger. It said that the celebrations would go from

January to April this year. That's a really big birthday party, I reckon.

Friday, 14th January

I met Dot at Bungaroo again today and she really wants to go to the Conference cos the people there are gunna try and stop making Aboriginal people work as domestic servants, like slaves. She said she could then do what she really wanted to do, which was be a school teacher. When I said I wanted to go too, she said there's no way cos we'd both end up getting sent away. I still wanna go though.

Saturday, 15th January

Ma and Pa B have gone out tonight to the Orpheum Theatre at North Sydney. I don't know what they've gone to see, but me and Sophie stayed home and one of Ma's friends came over to mind us. Sophie's out in the kitchen talking to her, but I'd rather be here writing to you. I wanted to go to the theatre too, but Ma said kids can't go. That's not true though cos Sophie told me she went last year to see 'Wanted, A Master' which was a show about a dog. She reckons it's a really nice place with cane chairs in the foyer and palm-tree plants and a cloakroom to hang your coat up so you don't have to scrunch it up on your lap. She reckons they have hygienic chairs with something called a carbolic block in them that destroys vermin and germs around the chair. There's meant to be a hat clip on the back of the chair where men can put their hats too. They'd need that cos men always wear hats. She reckons that after the show, trams and buses take people back to their homes, and that it's a really fun night out. Sophie wanted to go too, and when Ma B said kids couldn't go, Sophie said that she'd been before and Ma B went all funny-like. I reckon it's cos they didn't want to take me cos Aborigines can't sit in the same part and then they'd have to explain that to me. But I already know that stuff from Dot.

Monday, 17th January

The Sullivans have gone on holiday to Melbourne for two weeks so Dot has lots of time to herself and wants to do somethin' nearly every day if I can. I suppose we should make the most of the time while they're not spyin' on her all the time. But I feel bad lying to Ma all the time too but what other choice do I have?

Tuesday, 18th January

I met Dot at Bungaroo again today and she wanted to talk about the Conference and told me about all the people that are gunna be there. There's some famous Aborigines who will give speeches and she though it would be good for me to know who they are. She said there's a woman called Pearl Gibbs who was born at Botany Bay in 1901. That makes her 36. Geez, that's old, eh? She works with other people on getting Aborigines equal rights with white people. Other people like Jack Patten, who was born at a place called Cumm-er-oo-gunja but has lived at La Perouse for about eight years, she thinks. Dot reckons this Jack Patten bloke is really good at talking in public, and sometimes talks about Aboriginal rights and stuff down at the Domain in Sydney.

The other person Dot wanted me to know about was William Ferguson. Most people call him Bill. He's really old cos he was born in 1882. Geez, that's last century. Dot reckons he's a really good speaker too and helped set up the Aborigines Progressive Association with Jack Patten. It's this Association that's organising the big conference in Sydney.

Dot says the Association was only set up this year.

It has three main aims: to get citizenship rights for Aborigines; to get some Aborigines in the parliament, that means in government; and to get rid of the NSW Aborigines Protection Board. Dot says that's a big job, but she reckons that Aborigines are good fighters, even though we don't have guns.

Wednesday, 19th January

There was a new poster in Mr Johnson's shop today about the 150th birthday celebrations. It said 'Follow The Sun 1938' and has three people on it. I think a man, his wife and daughter. They was white people and had big white smiles that almost looked like they was painted teeth.

Friday, 21st January

I told Dot that I wanted to go to the Conference with her. She just said, 'No!' I was really shocked cos I thought she would want me to go. I asked her how she was getting there anyway. She's gunna get the train from Pymble to Town Hall and then walk to Australian Hall.

I didn't tell her, but I'm gunna act sick on 26th January and hope that Ma and Pa B let me stay home by myself. They wouldn't normally do that but all of their friends are goin' to the big parade and there'll be no-one to mind me, so they'll have to leave me. After the Burkes leave, I'm gunna take Sam's bike and ride to the train station and go in by myself. I reckon I can do that.

But what if Ma and Pa B won't let me stay home? I'm not even goin' to think like that cos it will make it happen. I gotta think positive, eh? I'm getting excited too, now.

Saturday, 22nd January

There was so much stuff in the newspaper today about Australia's birthday on the 26th. The papers say that it's a day to celebrate when the First Fleet landed at Botany Bay on 26th January, 1788. But there was no writin' in the paper about how Aborigines have died since these First Fleet boat people landed. I wonder why? Maybe they don't know all that stuff. Someone should tell those newspaper people all the true stuff so they can write about it too.

Sunday, 23rd January

Everyone at church today was talking about the Empire Games that are gunna start on 27th January (that's my birthday too), and lots of people were complaining about how expensive it was to go. I don't think we'll be going anyway. Ma and Pa B haven't said anything about it, but Pa B loves sports so he might go with his mates.

Monday, 24th January

Well it's nearly time for the big Conference, and I still don't really know what a conference is, but geez I'm getting excited about it. Dot and I met really quickly at Bungaroo today and she read the Sullivan's newspaper to me cos they're still away on holidays so she can read it. She gave me the story from the paper and I'm gonna write it in here so I can keep it forever.

> Three Aboriginal delegates from Victoria arrived in Sydney last night (Sunday 23rd January 1938) to attend the Day of Mourning Congress on Wednesday (Anniversary Day).
>
> They were Mr W Cooper, Secretary of the Australian Aborigines League, Mrs Marjorie Tucker, and D Nicholls, interstate Australian Rules player.
>
> The Congress will be held in the Australian Hall, Elizabeth Street. A Melbourne woman public speaker, Mrs Helen Baille, drove the

> delegates to Sydney in her car.
>
> Miss Baille said last night that the terrible conditions under which Aborigines lived in Australia had caused her to devote her time and energy and money to their cause.
>
> 'I have been in Central Australia and witnessed the degrading manner in which the Aborigines are treated by white people.'
>
> 'They are treated like many cattle: their women are debauched and the men flogged and killed.'

This story in the newspaper makes me wonder how the Aborigines live in the bush and the desert. It sounds really bad. Dot told me it was pretty bad for most Aborigines. I must be one of the ones who's got everythin' they should have with my food and school and stuff, eh? But maybe those other Aborigines got to stay with their real families and that's even more important.

PS: I had to look up the word 'debauched', and I think it might have to do with the kind of women Matron says wears lipstick.

Tuesday, 25th January

Oh I'm feelin' sick . . . only gammin—that's an Aboriginal word Dot taught me for 'just kidding'. I am just starting to practise being sick for tomorrow. I didn't eat all of my dinner to make it look like I wasn't feelin' well. We was having brussel sprouts and they'd probably make me sick anyhow. I got Sophie to feel my forehead and convinced her that I had a temperature so she's on my side, without even knowing it. She's not bad for a foster-sister. We get on heaps better now than when I first got here, eh?

Wednesday, January 26th—Australia Day

Well we did it. It was so much fun. I'm not even home yet, just sitting on the train, which is why my writin' is so bad. Hope you can read this.

I woke up and acted sick and, sure enough, Ma B said I'd have to stay in bed all day and that she'd stay with me. But you could see on her face that she was really sad about not being able to go. Pa B asked that if he locked the house up would I promise to stay in bed so that Ma B could go to the big parade. I said of course and so they did.

And then the plan went into action. After the Burkes

left, I got changed into my best dress and got on Sam's bike and rode to the train station, hoping no-one I knew saw me. My heart was poundin' cos I had to pedal really fast but also cos I knew I was really, really misbehaving this time. Ma and Pa B would be really angry if they knew. And what about Sam? He'd kill me if he saw me on his new bike, eh? It was really nice to ride too, with the breeze in my face and everything. It reminded me of when I used to ride the Sisters' bike back at Bomaderry.

I hid the bike behind some trees and hoped no-one took it. I don't think anyone saw me hide it there. At the train station I saw some people from school but no-one took any notice of me. I think they was all really excited about going into town for the parade. Then Mr Johnson from the shop saw me and said 'Hello' and I just said I can't talk cos Ma and Pa was waitin' for me just down a bit. He didn't even think I was lyin' or anything. I thought maybe Tambourine Talence should be an actress, eh?

I got on the train and followed all the instructions that Dot gave me and got off at Town Hall. There was people everywhere. I felt real strange doing this all by myself but I was so excited and my heart was beating really fast. You know when something good is about to happen but you're a bit nervous too?

I made it to Australian Hall at 148 Elizabeth Street by 10 o'clock and everyone was already standing outside for the big conference. It was runnin' late cos they had to stand out the front and wait for the other parade stuff to happen. I was lookin' around really worried cos now I was here and I was all by myself. I thought I must be crazy doing this. I'm just a kid. What if Dot didn't even come? Nah, she'd be there for sure.

I was lookin' around and there were other kids there too, some small ones, not babies though, just kids a bit younger than me with their parents. Everyone musta thought I was there with my parents too, cos no-one really said anything. Then all of a sudden, I saw Dot's hat, the really good one she wears on special occasions. And she turned around and saw me and had the biggest smile on her face. I ran to her and she gave me the biggest hug, like she hadn't seen me for ages. She said she knew I'd be there, but she was a bit worried about me travellin' by myself. I told her everything was fine and that I wouldn't have missed the meetin' for anything. Not even a big piece of Ma B's chocolate cake.

Then we just stood around and Dot was talkin' to all these people who were taller than me. Someone handed me a piece of paper with 'Aborigines Claim Citizenship Rights!' on it. It had Jack Patten's and William Ferguson's names on it. I was glad that Dot had told me about these people and about citizenship rights. I felt really clever that I knew what was going on. And I felt good cos I was there and the pamphlet said that 'Aborigines and persons of Aboriginal blood only are invited to attend'. That meant that no white people could go, and that I really was an Aborigine in the eyes of all these people. I liked that feelin'. And I didn't feel ashamed at all.

There were so many Aborigines there. I'd never seen so many in one place at one time. There musta been about a hundred people, and everyone was all dressed up and looked like they was goin' to church or something. I felt real good being amongst all the Aboriginal people. I didn't feel different like I do at school or at church. It wasn't like at the Home cos it was all Aboriginal people, and at Bomaderry there was Matron and the Sisters who was white. This was great with just Aborigines.

The meeting didn't start until 1.30 pm though and Mr Patten seemed angry when he was talkin'. He talked about how white people treated Aborigines very bad, and that we should have a voice. Like Dot told me before, Mr Patten said that the Aborigines Protection

Board of NSW should be abolished. That means got rid of. He also said that Aborigines are the old Australians and that whites are the new Australians. That's funny cos Johnny Jones says that Toni and all the 'other wogs' are new Australians. Bet he wouldn't realise that he's a new Australian too, eh?

Anyway, everyone seemed to be agreein' with everything he said and then he read from the pamphlet something on behalf of all the Aborigines at the meeting and even those Aborigines who couldn't make it, because it was a long way to travel to Sydney for most people.

I'm going to write what he read here because I think I should throw the pamphlet out in case Ma B finds it and wants to know where I got it from. I just hope she never reads my diary or I'll be in real big trouble. Luckily Sophie doesn't know where I hide you or I know she'd be reading you every night to see what I write about her.

Anyway the pamphlet says:

> We, representing the Aborigines of Australia, assembled in conference at the

> Australian Hall, Sydney, on the 26th day of January 1938, this being the 150th Anniversary of the Whiteman's seizure of our country, hereby make protest against the callous treatment of our people by the whiteman during the past 150 years, and we appeal to the Australian nation of today to make new laws for the education and care of Aborigines, and we ask for a new policy which will raise our people to full citizenship status and equality within the community.

I have to put you away now cos we're gunna get off the train in a minute. I'll finish the story when I get home. Geez, we're gunna have to really run to beat Ma and Pa home.

Later

I'm home again so I'll tell you that when we went in the Hall there were lots of speeches and sometimes I got bored listenin' all the time, but people would clap and cheer really loud and that would wake me up proper again. They was talking about citizenship rights for Aborigines. I saw that William Cooper man and Mr Ferguson. That Mr Ferguson was sayin' that there should be Aboriginal teachers and nurses to help Aboriginal people. I saw Dot smile cos she wants to be a teacher.

There was also the woman Dot told me about called Pearl Gibbs, who looked like a real smart lady. She even said that some of the people near Brewarrina didn't even have milk for their porridge. Geez, Sophie gives milk to her cat every night and these people don't even have milk for their porridge. That doesn't sound right, eh? Dot said Pearl Gibbs was a pea-picker and a unionist, which means someone who organises workers. She must be a 'Well Organised Girl' (WOG) too!

Another man who spoke was a Mr Foster, but I don't think he was related to Miss Foster at school. This man was from La Perouse. He was saying that alcohol was really bad for Aborigines but that white people had brought it with them to Australia when they first came here. Matron Rose said we should never drink alcohol, but she didn't say that the white man gave it to us to drink. I was really confused about this stuff.

There were lots of famous Aboriginal people there, so I was really surprised that I could stand so close to them. And I was lookin' real hard to find some of my real family. A couple of times I saw women I thought looked like what my mum used to look like. Lots of the women had nice hats on too, and some with gloves and fancy shoes with purses to match.

Dot introduced me to so many people and I just kept smiling in case one of them recognised me from back home with my other family, my Aboriginal one. Then all of a sudden I was sitting by myself while Dot went to the ladies' room. I was singin' that song Mum taught me that I couldn't remember the last coupla lines of and this lady next to me started singin' it too. She knew the whole song. I nearly fell off my chair cos I thought it must be my real mum or something. But it was this really nice lady who said she was from Griffith and knew the same song. And the last line was, 'I'll never love blue eyes no more'. She said she used to sing it to her kids too when they was little. I told her my mum used to sing it to me but I was from Cowra. She said that maybe she might know who my mum is. When Dot came back she got all my details from Dot and said she'd be in touch one day through Dot. What did she mean? Maybe she'll find my real mum for me and tell her I'm not dead, that I just have a different name and am living in the Hives and that I'm being an Aborigine too.

Maybe I'll get to see her again too, real soon. Imagine that? I can just picture what it will be like. I'll wear my yellow dress with the ribbons to match and I'll take my guitar so I can play that song I wrote her. I think she'd like that. I'll ask Dot when I see her next who that lady was and if she can find my mum. I just hope Ma and Pa B don't find out, cos they might think I don't like them or something and that's not true. I just want to see my real family again and tell them about what I've been doin' all this time.

I better put the light out now cos they'll all be home soon and I better seem like I was resting in bed all day. I'm gunna write a song for Dot tomorrow cos she's been real good to me, like an aunty. Oh, that's right, she's my tidda!

PS: Just a last thought. It seems like today meant really different things to white people and to Aborigines. For white people it seems it was like celebratin' progress, but for Aborigines it was about mournin' everything they've had to give up for white people's progress. I wonder if many white people understand where Aboriginal people are comin' from, eh?

Thursday, 27th January—My Birthday

It's my birthday today but I'm more excited about what I did yesterday than turnin' eleven years old. And even though the family gave me a birthday cake with candles and some really nice presents, everyone was still talkin' about the parade and stuff.

Sophie reckons there was about a million people to watch the parade and that there were pretty banners in the city with brass bands and marching girls and everything. She even said there were Aborigines there too, and that at the Town Hall there was a play where some white school boys were painted brown and danced around pretending to be Aborigines. I bet they was tryin' to dance a corroboree like Dot talked about. But Sophie says that the music was like opera or something, but that doesn't sound right, eh? You can't make opera music with a didjeridoo and clap-sticks, can ya?

Ma B reckons that down at Farm Cove they showed what it was like when the boats first arrived from England with Aborigines waiting on the shore. She said everyone cheered at the Aborigines holding sticks. Dot says they're called nullu nullus. These must be the Aborigines from the Menindee Reserve that Dot told me about. They had to bring these Aborigines in on a truck to dance on the sand and act stuff out cos the Aborigines in the city didn't want any part of it. Dot said it was not right the way the white people invaded Australia in the first place. They shouldn't be doin' it again.

Sophie said she liked the big navy boats in Sydney Harbour cos they had lots of good-looking sailors on 'em. I said I only liked Joey from singing group and I'd marry him one day and not some sailor. Sam said I was dreamin', but he's just jelly bean cos he's not got any girls that like him. I don't reckon any boys like him either!

I listened to Sophie talk about the day she had but I didn't care. I just wanted to think about what I did and all the people I met. And maybe I might be able to find my mother, or maybe she'll find me first. That would be the best birthday present in the world, eh?

You know what I really liked about yesterday? That everyone was really proud to be Aboriginal, not like school where people make you feel ashamed for being Aboriginal. Everyone was saying they was Aboriginal and they was really proud. That made me proud too. I'm not gunna let people tell me no more that I have to be a white person, cos that's not being true. I reckon I'll even tell Ma B cos if she really does love me like a daughter then she has to love who I am. Right? And I'm Aboriginal and I'm proud.

HISTORICAL NOTE

In New South Wales there was a government body called the Aborigines Protection Board which was set up in 1883. In 1940 the Aborigine Protection Board changed its name to the Aborigines Welfare Board and they controlled Aborigines until 1969. There was one of these kinds of Boards in every state and territory of Australia. These boards thought that the only way Aboriginal children would have a good life was to take them away from their families to be raised in white families or orphanages, even though their real parents didn't want them to go.

The Government also had lots of policies about Aborigines and these were called the NSW Aborigines Protection Act(s) between 1909 and 1943. These Acts meant that the Aborigines Protection Board had control of Aboriginal peoples' lives including how much money they got paid and who controlled that money, where Aborigines could travel around the country, where Aboriginal children went to school, and even whether Aboriginal people could speak their languages, dance their dances and practise their religions. It was the Aborigines Protection Acts that said it was all right to remove, steal, or take Aboriginal children from their Aboriginal families and make them live somewhere else without telling the children why or letting them know where their families were.

Between 1915 and 1939, any station manager or policeman was also allowed to take Aboriginal children from their real families if he believed the children would be better off somewhere else, that is, with a white family, in a home or a training institution.

The Government also said that Aborigines with mixed blood, that means someone who had one Aboriginal and one non-Aboriginal parent, could be given an Exemption Certificate. This certificate was sometimes called a 'Dog Tag' and meant that anyone who received one was not to be treated like an Aborigine any more, but they also had to pretend they were white, and weren't allowed to have anything to do with their families.

The Government thought that the more of these certificates they handed out, the more Aboriginal people were becoming white and they thought that was a good thing. This is one way they assimilated Aboriginal people. 'Assimilate' is the big 'A' word that Dot told Mary, and it was supposed to make Aboriginal people become more like white people, and meant they had to have white religions, educations and lifestyles.

The church missionaries and schools were in charge of

assimilating Aboriginal people into the white world, and money was given from the church and the government to help this happen. In the 1950s, the Government made a statement to say that they wanted Aborigines to have the same rights and privileges as white people, but they still didn't have the right to vote, didn't get equal wages for the same work they did as white people, and the Government was still taking one in seven Aboriginal children from their real parents. Only one in three hundred white children were removed from their families.

The policies of protection and assimilation are the reason that so many Aboriginal families, communities and society suffered almost complete destruction. The children who were taken from their parents became known as the 'Stolen Generations' and number between 15 000 and 20 000 in NSW alone. Even during the First World War, when between 400 and 500 Aboriginal people served as soldiers, the Government still took Aboriginal children from their families, including children whose fathers were fighting overseas at the time.

As late as 1957 the Aborigines Welfare Board advertised for white families to adopt Aboriginal children, so it didn't cost the Government as much to run homes. These foster families often treated the children well, but they always knew that their job was to make the Aboriginal child fit into white society better.

The time of Mary's story was before the policies of fostering Aboriginal children were introduced, and most of the fair-skinned children were sent to mainstream homes like Burnside, where it would be easier to foster them out because no-one would know they were Aboriginal. White families sometimes didn't know they were adopting an Aboriginal child from these homes. But, people would know that the children from Bomaderry were Aboriginal and therefore no-one would want them.

Bomaderry Aboriginal Children's Home was opened in 1908 to look after seven Aboriginal children. Up to forty-seven Aboriginal children stayed there at any one time. The Home was on the Bomaderry Aboriginal Mission and was managed by the United Aborigines mission until 1988.

The Cootamundra Aboriginal Girls' Home opened in 1911 and the Kinchela Aboriginal Boys' Home in Kempsey in 1924. These were just two training homes set up specifically to train Aboriginal children to become domestic servants or farm labourers.

In 1980, Link-Up (NSW) Aboriginal Corporation was set up to help children and parents find each other.

RESOURCES

Anita Heiss used a large number of references for this book. One set were general resources to give background of the time. For example, books about cars of the period, about Sydney suburbs, the Sydney markets, and about the early history of the Ku-ring-gai area were consulted. These general references are too many to quote here.

Another set of references were those used to gain details of what happened to Aboriginal children during this time in order that the author could create the story of Mary Talence. Again, a large number of books was consulted, some of which are listed here.

Bird, Carmel ed, *The Stolen Children: their stories*, Random House, Milsons Point, NSW, 1998.

Bringing Them Home: A guide to the findings and recommendations of the National Enquiry into the separation of Aboriginal and Torres Strait Islander children from their families, HEREOC, 1992.

Edwards, Coral and Read, Peter eds, *The Lost Children: 13 Australians taken from their Aboriginal families tell of the struggle to find their natural parents*, Doubleday, Moorebank, NSW, 1989.

Haebich, Anna, *Broken Circles: Fragmenting Indigenous Families 1800–2000,* Fremantle Arts Centre Press, South Fremantle, 2000.

Read, Peter, 'Don't Turn Your Back on Me: A Bibliographical Review of the Literature of the Stolen Generations', *Aboriginal Law Bulletin,* Vol 3, No 73, April 1995, pp 22–24

Read, Peter, *The Stolen Generations: The Removal of Aboriginal children in New South Wales 1883 to 1969,* NSW Department of Aboriginal Affairs, 1998.

Read, Peter, *A Rape of the Soul So Profound,* Allen and Unwin, St Leonards, 1999.

'Securing the Truth: NSW Government Submission to the Human Rights Commission Inquiry into the Separation of Aboriginal and Torres Strait Islander Children from their Families', NSW Department of Aboriginal Affairs, 1998.

Wilson, Tikka Jan, 'Link-Up: In the Best Interest of the Child? the present effects of separating New South Wales Aboriginal Children from their families and recommendations to redress the ongoing harm of separation', Link-Up (NSW) Aboriginal Corporation Final Submission to the Human Rights and Equal Opportunity Commission National Inquiry into the Separation of Aboriginal and Torres Strait Islander Children from their Families, 1996.

ACKNOWLEDGMENTS

I'd like to thank the staff at the Australian Institute of Aboriginal and Torres Strait Islander Studies and the Aboriginal Research and Resource Centre (UNSW). Special thanks to Eileen Stevens, Barbara Nicholson and Link-Up (NSW) for their time, experience and wisdom. Thanks to Libby Gleeson, Rosie Scott and Linda Jaivin for proofing and unconditional friendship and support.

The photograph on the front cover is of Leila Clarke from Walgett NSW, taken around 1948. Leila had two half-sisters removed from her family before she had a chance to meet them, and they never came back. She later married Arthur Murray and became a committed advocate for Aboriginal rights, and as Leila Murray became known for her role in calling for the Royal Commission into Aboriginal Deaths in Custody as well as raising eleven children of her own.

ANITA HEISS

Dr Anita Heiss is from the Wiradjuri nation of central New South Wales, but was born and bred in Sydney. She is an author, poet, activist and social commentator. Anita is the author of three adult novels: *Not Meeting Mr Right, Avoiding Mr Right* and *Manhattan Dreaming* (all through Bantam Australia). She has published a non-fiction text *Dhuuluu Yala (talk straight): publishing Aboriginal literature* (Aboriginal Studies Press, 2003, a collection of social observations titled *I'm not racist, but...* and co-edited with Peter Minter *The Macquarie PEN Anthology of Aboriginal Literature*. In 2003 Anita was awarded the Australian Society of Authors Medal (Under 35s) for her contribution to Australian literary and community life, and in 2004 was listed on the *Bulletin*/Microsoft 'Smart 100'. Anita is currently working as a full-time writer and Adjunct Associate Professor at the University of Western Sydney attached to the Badanami Centre for Indigenous Education. Anita volunteers her time as a Books in Homes Ambassador and as an Indigenous Literacy Day Ambassador. She is currently a Board Member of the National Aboriginal Sporting Chance Academy. Her web site is *www.anitaheiss.com*.

Also Available:

North-east Victoria, 1879. Jamie Ross and his older sister Ellen are alone in the world after the death of their father. Determined to make their fortune, they head to Beechworth—and straight into the midst of the search for Ned Kelly, the most notorious bushranger of all time. A chance encounter will bring Jamie closer to the outlaw than he could ever have imagined.

Also Available:

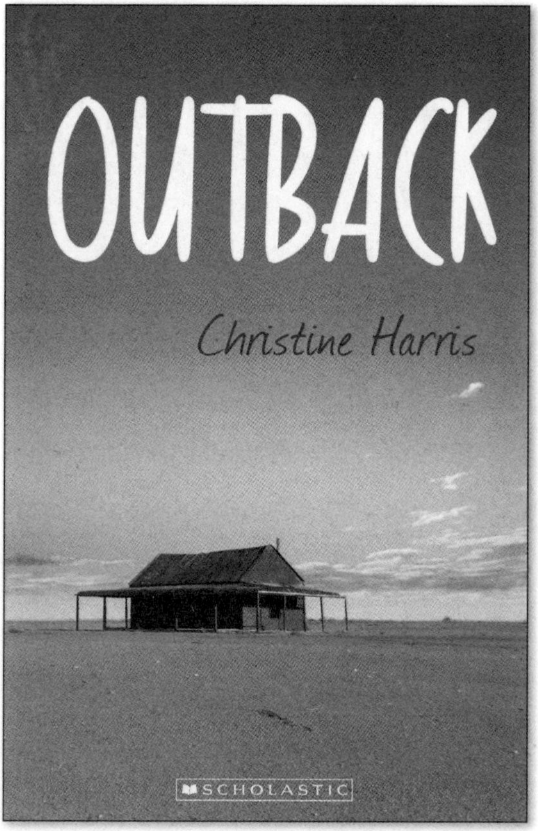

Jimmy Porter has moved to the middle of nowhere. His uncle's family live in a wattle-and-daub hut, days' walk away from even the nearest neighbour.

Life in 1927 in the outback is tough—but the people who live there can cope with just about anything.

But when disaster strikes, how can they get help?

Also Available:

It's 1969. The sixties are still swinging, man is headed for the moon and surfing-mad Davey is out to win the inaugural Newcastle Under-14 Surf Championship. But when his older brother Tom is drafted to fight in the army, the far-away war in Vietnam creates waves very close to home.

Also Available:

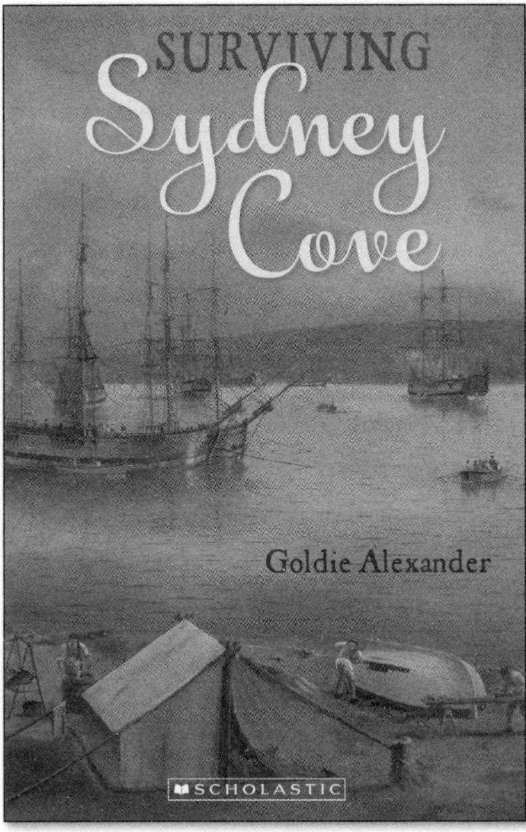

Lizzie Harvey is a convict, sent to Sydney Cove with the First Fleet. Starving and overworked, she can barely find time to dream about the way things used to be, let alone write in her diary. But write she must, in the hope that she will one day be able to share her story with the family she has left behind.